ELEMENTS
OF RADIOBIOLOGY

ELEMENTS
OF
RADIOBIOLOGY

By

JOSEPH SELMAN

M.D., F.A.C.R., F.A.C.P.

Director, School of Radiologic Technology
Tyler Junior College

Attending Radiologist, Mother Frances Hospital

Attending Radiologist, Medical Center Hospital

Consultant in Radiotherapy
University of Texas Health Center at Tyler, Texas

CHARLES C THOMAS • PUBLISHER

Springfield • Illinois • U.S.A.

Published and Distributed Throughout the World by
CHARLES C THOMAS • PUBLISHER
2600 South First Street
Springfield, Illinois 62717 U.S.A.

*With THOMAS BOOKS careful attention is given to all details of
manufacturing and design. It is the Publisher's desire to present books that are
satisfactory as to their physical qualities and artistic possibilities and
appropriate for their particular use. THOMAS BOOKS will be true to those
laws of quality that assure a good name and good will.*

Library of Congress Cataloging in Publication Data

Selman, Joseph.
    Elements of radiobiology.

    Includes bibliographical references and index.
    1. Radiobiology. I. Title. [DNLM: 1. Radio-
biology. WN 610 S467e]
QH652.S44 1983 612'.014485 82-10536
ISBN 0-398-04753-7

Printed in the United States of America
SC–R–1

PREFACE

For a number of years, the author has given a course in radiobiology to second year students in radiologic technology, as well as an occasional lecture to radiology residents. More recently, radiobiology has become a mandatory subject in the curriculum of approved radiologic technology schools.

In this book, the author has expanded and thoroughly updated his lecture notes so as to conform to the required curriculum. Since the book is directed both to technology students and beginning radiology residents, emphasis has been placed on the essentials of radiobiology.

After a brief historical introduction, basic physics and cellular biology pertinent to radiobiology are treated separately, followed by chapters on their interplay: modes of action of x and gamma rays, response of cells and tissues, cellular radiosensitivity, and factors affecting cell response to ionizing radiation.

Because of the growing interest in radiation hazards, fully one-third of the book deals with this important subject. Thus, it includes chapters on whole body effects, hazards to embryo and fetus, late effects on body tissues, genetic effects, and population exposure (health physics).

Radiobiology has become such an integral part of irradiation therapy that an introduction to radiation oncology and an overview of available radiation modalities have been included. Finally, there is a brief discussion of radiotherapy: equipment, goals, planning, and terminology.

The author gratefully acknowledges the kindness of various authors and publishers for permission to use their published material. Credits are shown in an appropriate manner where borrowed material has been used.

The illustrations were skillfully drawn, in the original, by Howard Marlin. His excellent work is greatly appreciated.

Finally, the author wishes to thank the staff of Charles C Thomas, Publisher, for their sound advice, cooperation, and diligence in the various stages of publication.

Joseph Selman, M.D.
Tyler, Texas

CONTENTS

Brachytherapy
   Interstitial therapy
   Systemic therapy with radionuclides
Radiotherapy
   Goals of radiotherapy
   The planning process
   Terminology in radiotherapy
   Observation of the patient

ELEMENTS
OF RADIOBIOLOGY

Chapter I

INTRODUCTION

T hree remarkable events, in rapid succession, ushered in the era of radiology. In 1895 Wilhelm Conrad Roentgen discovered x rays through their ability to excite fluorescence in a barium platinocyanide screen. The following year Henri Becquerel first observed the radiations emitted by a uranium-containing mineral, a phenomenon later called *radioactivity* by Marie Curie. Then, in 1898, Marie and Pierre Curie announced their discovery of radium. In each instance the important product was *ionizing radiation.*

The biologic effects of such radiation soon became apparent when Becquerel noticed a skin reaction—reddening and irritation— accidentally induced by radium he had been carrying in a tube in his vest pocket. Later, Pierre Curie deliberately exposed a small area of his own skin to radium and closely observed the consequent radiation effects.

History of Radiobiology

We may define radiobiology as the branch of science that deals with the modes of action and the effects of ionizing radiation on living matter. This important discipline has contributed in no small measure to our understanding of the factors involved in population exposure to radiation, as well as tumor response to irradiation, although radiotherapy still remains largely empirical (i.e., based on practical experience).

We can date the beginning of experimental radiobiology to Bergonié and Tribondeau who, in 1904, exposed rabbits' testes to x rays. Their observations of radiation effects on testicular reproductive cells are embodied in the law that bears their names. This law simply relates radiation sensitivity or responsiveness of a tissue, to the fraction of cells that are actually dividing (i.e., in mitosis) or have the potential to divide in the future. While this

3

law holds generally and has been verified in a great many different types of tissue, it is subject to certain exceptions that will be brought out later.

In the 1920s it was already widely accepted that the *biologic effects* of radiation result from the ionization it produces in tissues. It soon became apparent that two different processes are involved: (1) direct effects by ionization along charged particle tracks and (2) indirect effects by free radicals and other entities that diffuse (spread) away from the ionization tracks. This second process, *activation of water*, received a great deal of attention in the 1930s by H. Fricke (1936), and later by D. E. Lea (1947) and L. H. Gray (1953).

Another important step in the advance of radiobiology was the discovery by Thoday and Read (1947) that oxygen increases the frequency of chromosome breaks over and above that predicted from the indirect ionizing effect of radiation such as x and gamma rays. It turns out that oxygen enhances the indirect effect exerted by free radicals. This, the *oxygen effect*, was also studied by Gray, an outstanding pioneer radiobiologist.

In 1956 Puck and associates reported the first successful culture of mammalian cells in artificial media, much like bacteria, thereby opening the door to large scale investigation of cellular response to radiation under controlled laboratory conditions. They used cells derived from a carcinoma of the uterine cervix in a patient named Helen Lane; these cells are called HeLa cells. Puck and associates exposed such cells to various doses of radiation and plotted cell survival curves. Based on curves of this kind, Elkind in the 1950s was able to show that cells can recover from sublethal (less than fatal) doses of radiation.

It should be noted that a definite interrelation exists among the three disciplines of radiation physics, biology, and radiobiology. *Radiation physics* deals with the spread of energy through space and its absorption in matter. *Biology* comprises all the available knowledge about living organisms. As noted at the outset, *radiobiology* includes the interaction of ionizing radiation with living systems. Later we shall describe the salient features of experimental radiobiology and show how they relate to radiotherapy.

Chapter II

PHYSICAL PROPERTIES OF X-
AND GAMMA-RAY BEAMS

Nature of Photon Radiation

Photon beams, whether x rays or gamma rays, consist of electromagnetic radiation characterized by high frequency and short wavelength. They belong to the general electromagnetic spectrum (i.e., range of frequencies) as illustrated in Figure 2.01.

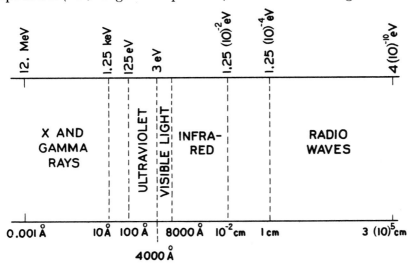

$$1 \overset{\circ}{A} = \frac{1}{100,000,000} \; cm = 10^{-8} \, cm = 10^{-10} M$$

Figure 2.01. Electromagnetic spectrum. Useful range of wavelengths and energies is as follows:

Diagnostic x rays	0.1 to 1 Å (124 to 12.4 kV)
Cobalt 60 gamma rays	0.01 Å (1.25 MeV)
Linac; betatron	0.0005 to 0.002 Å (25 MV to 6 MV)

5

X rays arise in association with the following processes:

1. Change in velocity (i.e., speed or direction) of high-speed electrons.
2. Transition of electrons between atomic shells, from higher to lower energy levels.
3. Rapid oscillation (vibration) of electrons.

Gamma rays—physically identical to x rays—originate in the nuclei of certain radionuclides during radioactive transformation (decay).

All electromagnetic radiation consists of simultaneous electric and magnetic waves that proceed through space in step with each other as shown in Figure 2.02. Ordinarily, only one of these waves

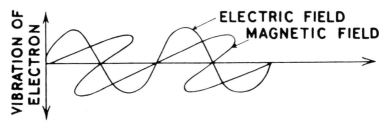

Figure 2.02. Electromagnetic wave form. The electrical and magnetic components of an electromagnetic wave oscillate (vibrate) in mutually perpendicular planes.

is used by way of illustration, as in Figure 2.03.

Electromagnetic waves have the properties of *frequency*, symbolized by the Greek letter ν (*nu*) and *wavelength*, symbolized by the Greek letter λ (lambda). Frequency designates the number of vibra-

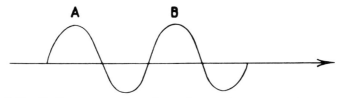

Figure 2.03. Ordinary representation of an electromagnetic wave, comprising only one component. The distance between two successive peaks, such as A and B, is the wavelength. The part of the wave between two corresponding points such as A and B is one cycle, and the number of cycles per second is the frequency.

tions or cycles per second for a particular electromagnetic wave, measured in hertz (1Hz = 1 cycle/sec). The distance between two successive crests in the wave such as *AB* in Figure 2.03 is the wavelength.

The frequency and wavelength of any given electromagnetic wave have a reciprocal relationship. Since the frequency determines the number of waves per second passing a given point in space, the speed of the wave equals the frequency times the wavelength, symbolized as follows:

$$c = v\lambda \qquad (1)$$

where *c*, the speed of light, is a universal constant in air or in a vacuum. From equation (1) you can see that since *c* is constant for all electromagnetic radiation, *an increase in* v *must be accompanied by a corresponding fractional decrease in* λ. For example, if v is doubled, λ is halved; if v is tripled, λ is reduced to $\frac{1}{3}$, since their product must always equal *c*, which represents 3×10^8 meters per sec in air or vacuum.

Owing to the extremely short wavelength of x and gamma rays, much smaller units of measurement are more convenient than those in everyday use. These include the angstrom (Å) having a value of 10^{-10} meters (m) or 1/10,000,000,000 m, and the micron or micrometer having a value of 10^{-6} m or 1/1,000,000 m.

Curiously enough, x and gamma rays exist not only as electromagnetic waves but also as extremely minute (tiny) bits of energy called *quanta* (plural of *quantum*) or *photons*. The energy of *E* of such a quantum is defined by:

$$E = hv \qquad (2)$$

where *h* is Planck's constant (6.625×10^{-27} erg-sec) and v is the frequency of the electromagnetic wave associated with the quantum. Thus, the *energy of a quantum or photon is directly proportional to its frequency*. We shall find that the interactions of x and gamma rays with matter can be explained only with reference to the quantum aspect of radiation. These *simultaneous wave and quantum properties* express the *dual nature* of electromagnetic radiation.

On the basis of this concept, known as the *quantum theory*, an x- or gamma-ray beam consists of showers of photons or quanta (bits of energy) that carry no charge. Furthermore, *they travel with the same, constant speed in air or vacuum regardless of their energy.* Thus, x and gamma rays (as well as light) have a dual nature, behaving as both waves and particles.

Since, from equation (2), the energy of a photon is proportional to its frequency, and from equation (1), the frequency times the wavelength is constant (speed of light *c*), highly energetic (very penetrating) photons have high frequency and short wavelength; conversely, low energy (less penetrating) photons have low frequency and long wavelength. This relationship is shown in Figure 2.04.

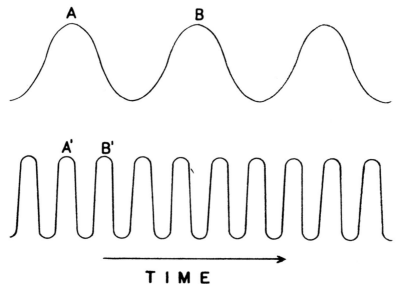

Figure 2.04. Two electromagnetic waves. The upper and lower wave trains differ in wavelength (compare AB to A′B′). The lower wave has a shorter wavelength, and therefore more peaks (or cycles) per given time interval, than does the upper wave because their velocities are identical. So the wave with the shorter wavelength has the greater frequency or number of cycles per second (Hz).

Quantity of Photon Radiation

By the term *quantity* we mean the amount based on a particular property of ionizing radiation, namely, its *ability to ionize air*. You should recall that ionization refers to the separation of atoms or molecules into positively and negatively charged ion pairs by removal or addition of electrons. The ionizing ability of radiation has been in use only for the last fifty years as a measure of quantity.

The early radiotherapists estimated radiation quantity by the reddening effect (erythema) produced in the skin of the forearm. Thus, the threshold erythema dose (TED) was defined as that quantity of radiation in a particular x-ray beam that caused a barely perceptible reddening of the skin in 48 hours. Obviously, the TED varied with the observer, the complexion of the subject, and other factors, so it was eventually abandoned as a measure of quantity.

We now have three interrelated ways of expressing radiation quantity: (1) exposure, (2) absorbed dose, and (3) dose equivalent. The first two will now be described, but the third will be considered later since it relates to radiation protection only.

Exposure

The first modern unit of radiation quantity, officially called *exposure*, was based on ionization produced in air by a photon beam. Its unit, the roentgen (R), is defined by:

$$X = \Delta Q/\Delta m \qquad (3)$$

where X is exposure in R, and ΔQ is the sum of all the electric charges on all the ions of one sign (+ or −) produced in air when all the electrons released by photons (x and gamma rays) in a mass of air, Δm, are completely stopped in air; that is, the released electrons have produced as many ions as they can before losing all their energy. Note that the definition implies two steps in the process: (1) photons first release electrons by interactions with atoms in the air mass by processes to be described in the next chapter, and (2) these released primary electrons in turn interact with other atoms to produce ion pairs—the total charge carried by all the ions of one sign being the measure of radiation exposure. On this basis,

$$1 \ R = 2.58 \times 10^{-4} \ coulomb/kg \ air$$

the coulomb being the unit of electric charge.

The roentgen is scheduled to be discontinued as a unit in the new International System of Units (S.I.) and is to be replaced by

$$1\ exposure\ unit\ =\ 1\ coulomb/kg$$

However, because of its wide prevalence in medical radiology, we shall continue to use the roentgen at least for the present.

The exposure per unit time such as *R per min* (R/min) is called the exposure rate:

$$\underset{(R/min)}{exposure\ rate} = \frac{exposure}{time\ in\ min} \qquad (4)$$

The total exposure in R is obtained by multiplying the exposure rate by the exposure time. Rearranging equation (4),

$$\underset{(R)}{exposure} = \underset{(R/min)}{exposure\ rate} \times \underset{(min)}{time} \qquad (5)$$

You can see that the physical units in equations are handled in the same way as numbers; thus, they can be multiplied, divided, or raised to a power.

Absorbed Dose

Another concept of radiation quantity is the *absorbed dose.* This specifies the amount of energy released in the tissues and absorbed there as the result of a given radiation exposure. Note that exposure in R tells us the amount of radiation, in terms of its ionizing ability, to which matter (tissues included) has been subjected; it states nothing directly about the amount of radiation absorbed. Furthermore, the exposure in R applies only to photon radiation having an energy up to 3 MeV on account of limitations inherent in the measuring systems. On the other hand, absorbed dose applies to any type of ionizing radiation, regardless of its energy.

The unit of absorbed dose is the *rad,* defined as an absorption of 100 ergs per gram of matter. (The erg is an extremely small unit of energy.) The relationship between exposure and absorbed dose will be discussed further in Chapter III.

Measurement of Exposure

We shall deal briefly with this subject, as it is amply covered in standard textbooks on radiologic physics (see Bibliography). For maximum precision, exposure measurement requires a special instrument known as a *standard free-air ionization chamber*, which collects and measures the ions produced by a beam in a well-defined small volume of air. This bears out what has already been stated: *the unit of radiation exposure is based on ionization in air.*

Because the standard free-air ionization chamber is a delicate laboratory device that requires meticulous technic, it does not lend itself to practical calibration of equipment in the radiology department. Instead, we have available a number of *secondary R-meters* that are much more convenient, and still sufficiently precise for measuring R output. All have in common a thimble-type capacitor serving as an ionization chamber (see Figure 2.05).

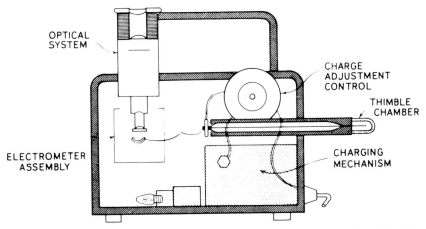

Figure 2.05. Diagram of a Victoreen-R-meter based on data furnished by the Victoreen Instrument Corporation.

They differ in the type of electrometer used to measure the collected ions resulting from the entrance of radiation into the thimble chamber. The following instruments are available at present:

Victoreen Meter. This is a string-type electrometer, which is separable from the thimble chamber assembly. The latter is exposed to the radiation beam under specified conditions that are easily fulfilled. It is then inserted in the electrometer in another room for measurement.

Baldwin-Farmer Meter. This is a string-type electrometer connected by a long, shielded cable to the thimble chamber, an arrangement that obviates carrying the thimble chamber back and forth from the radiation source to the electrometer outside.

Philips Universal Dosemeter. The dosemeter uses a vibrating reed type of electrometer connected to the thimble chamber by a long, shielded cable.

Solid State Electrometer. The most modern type, this uses transistorized circuitry. Because of its sturdiness and reliability, the solid state unit promises to replace the other types of R meters.

Whatever type of R meter is selected, it must be calibrated at frequent intervals by an approved laboratory. Otherwise, malfunction of the instrument could result in serious errors, especially in the measurement of R output of therapy machines.

Radiation Quality

By the term *quality* of a photon beam we mean its *penetrating ability*. More specifically, in *radiography* the penetrating ability of a particular beam may be regarded as the amount of radiation reaching the film relative to the amount entering the body; or, expressed in another way, the ratio of the exit exposure to the entrance exposure (see Figure 2.06A). In *radiotherapy*, the penetrating ability of a beam refers to the percent depth dose, that is, the ratio of the dose at a specified depth in the body to the maximum dose where the beam enters the body; thus, the greater the penetrating ability of a beam, the greater will be the dose of radiation reaching a specified depth, relative to the maximum dose near or at the entrance surface (see Figure 2.06B).

For a given body of matter, the penetrating ability depends on beam energy, and we have just seen from equation (2) that the

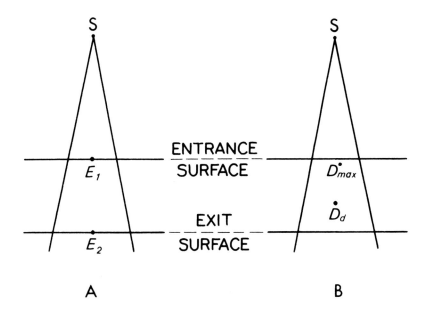

Figure 2.06. Concepts of penetrating ability and percent depth dose. In *A* the penetrating ability may be represented in terms of percent by the expression $100E_2/E_1$ that is, 100 (exit exposure ÷ entrance exposure).

In *B* the percent depth dose is $100D_d/D_{max}$ that is, 100 (dose at given depth ÷ given dose).

energy of a photon is proportional to its frequency. Consequently, a more penetrating beam with higher energy contains photons of high frequency (or short wavelength). Keep in mind that all photons, regardless of their energy, have the *same speed* in air or vacuum.

In general, the energy of all photon beams is expressed in units that are multiples of *volts*. However, this requires modification according to the following categories:

Radiographic X-ray Beams. Generated at maximum potentials ranging from about 30 to 120 kV across an x-ray tube, such beams are *polyenergetic* or *heterogeneous*, containing as they do photons in a *spectrum* or *energy range*. This results from (a) periodic fluctuation of kV across the tube, from zero to a peak value (see Figure 2.07),

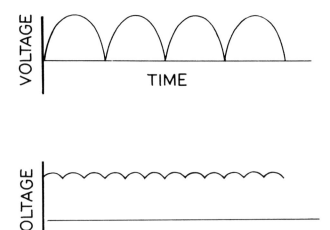

Figure 2.07. Voltage variation across an x-ray tube. In the upper diagram we see the wide fluctuation in voltage as a function of time in a *single phase* full-wave rectified circuit. In the lower diagram the voltage is more nearly constant, showing a slight "ripple" as in *three-phase* x-ray equipment.

(b) production of brems radiation in a continuous spectrum, and (c) production of characteristic radiation in discontinuous peaks. Such beams require about 3 mm total aluminum filtration (or equivalent) to remove the lower energy (i.e., lower kV) components that cannot penetrate the body sufficiently to be image-producing. Their energy may be specified in terms of peak kilovoltage, that is, kV_p. However, we usually omit reference to "peak" in the radiographic range and simply say *kV*.

Orthovoltage X-ray Beams. Although these ordinarily include therapy beams in the energy range of 200 to 300 kV_p, we may extend this range down to 100-kV therapy. Such beams are polyenergetic but differ from radiographic beams in that they require various degrees of filtration, depending how they are to be used. So, we cannot express the quality of orthovoltage radiation by a single kV_p value because it does not adequately reflect beam quality in radiotherapy. We must, therefore, resort to another way of specifying orthovoltage beam quality, namely, the *half-value layer (HVL)*. By *definition, HVL simply states the thickness of a particular filter material that reduces the exposure rate of a beam to exactly one-half its initial value.* For example, a series of exposure-rate measurements are made after inserting filters of different suitable thicknesses in the selected beam (see Table 2.01).

TABLE 2.01

EXPOSURE RATES WITH ADDED COPPER FILTERS
OF VARIOUS THICKNESSES.   kV, mA,
AND DISTANCE ARE CONSTANT THROUGHOUT.

Added Copper Filtration	Exposure Rate
mm	R/min
0	80
0.25	40
0.50	29.6
1.0	20
1.5	15.2

These data are then plotted graphically, with exposure rate (say, R per min) on the vertical axis, and filter thickness on the horizontal axis (see Figure 2.08). From the resulting curve we find that for this particular beam a 0.25 mm copper (Cu) filter reduces the initial exposure rate to one-half its initial value. We may therefore state the HVL of this beam to be 0.25 mm Cu. Note by way of emphasis that HVL does *not* specify directly how much filtration is to be used with a particular beam; it simply serves as a convenient measure of *beam quality* or *penetrating ability.* However, we customarily state *both the HVL and the* kV_p *in specifying orthovoltage beam quality.* In practice, orthovoltage beams for "deep" therapy have ordinarily been in the range of about 200 to 300 kV with sufficent initial filtration to provide a beam with HVL of 1.0 to 2.0 mm Cu.

Megavoltage X-ray Beams. Such beams, generated by application of a million or more volts across specially designed x-ray tubes, are usually specified in megavolts or MV (1 MV = 1 million volts). For example, the x-ray beam put out by a certain linear accelerator (linac) may be stated as having an energy of 6 MV.

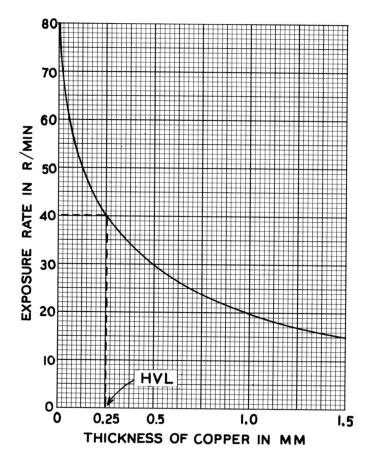

Figure 2.08. Concept of half-value layer (HVL) based on an absorption curve of 220-kV x rays in copper filters of increasing thickness. Assuming an exposure rate of 80 R/min of the initial beam, we find that as we add copper, the exposure rate declines along the curve. Since the HVL is that thickness of a particular material (in this case copper) that reduces the exposure rate by one-half, we locate 40 (i.e., one-half the initial exposure rate) on the vertical axis and trace horizontally to the curve. From the point of intersection we trace vertically downward to the horizontal axis where we find 0.25 mm Cu as the HVL or quality of this particular beam.

Gamma-ray Beams. Because radionuclides emit gamma rays in one or more groups of monoenergetic (single energy) photons, we specify their energy in multiples of electron volts (eV). An eV is the energy an electron gets when it falls through a potential difference of 1 volt. Some radionuclides emit more than one set of monoenergetic photons. For example, cobalt 60 (^{60}Co) gives off two different groups of gamma rays, one with an energy of 1.17 MeV (million electron volts), and the other with an energy of 1.33 MeV, the average being 1.25 MeV. On the other hand, cesium 137 (^{137}Cs) emits only 0.66 MeV gamma rays. Although the HVLs of various gamma-ray beams have been determined, we usually state only the energy in specifying beam quality just as with megavoltage x-ray beams, except that in the latter we express the energy in MV rather than MeV.

Chapter III

THE BEHAVIOR OF X AND
GAMMA RAYS IN MATTER

ENERGY TRANSFER BY PHOTONS

On passing through body tissues (or any kind of matter) x and gamma rays lose energy (*not* speed). Such energy release by photons in matter occurs in *two steps*: (1) the photons first interact with atoms, liberating orbital electrons, and (2) these freed electrons, called *primary electrons*, then move through the body causing atomic ionization and excitation as they interact with other orbital electrons. Thus, photon radiation ionizes *indirectly* in that it first sets in motion orbital electrons in very small numbers, each of which then releases many thousands of secondary electrons, thereby *greatly multiplying the ionizing effect*. This two-step process underlies the physical basis of radiobiology and will now be described in some detail.

STEP 1—Photon Interactions with Matter

Let us now review the *first step*, that is, the liberation of primary electrons by photons. This occurs by one or more processes (photoelectric; Compton; pair production) depending on the quality of the radiation and the atomic number of the atoms with which it interacts. The matter in which photons deposit energy may be regarded collectively as the *absorber*.

Before describing the main types of interactions between photons and atoms we must understand that the electron shells within an atom have progressively higher energy levels the farther a particular shell is from the nucleus. As shown in Figure 3.01, the K shell represents the lowest energy level, the L shell the next higher one,

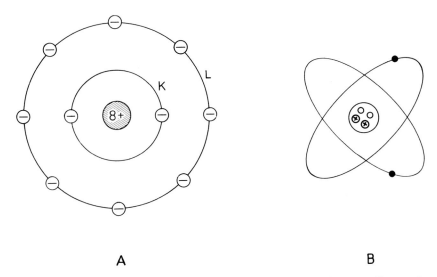

A   B

Figure 3.01. In *A* is shown a greatly simplified atomic model according to the Bohr theory. Only the two innermost shells or energy levels are shown, situated around the central nucleus. The *K* shell can hold no more than two electrons, and the *L* shell no more than eight. Actually, the electrons do not "chase" each other around the same path; as shown in *B*, for example, each of the *K* electrons has its own orbit in which it revolves very rapidly around the nucleus, at the same time spinning on its axis. Since each *K* electron has the same energy level with respect to the nucleus, both electrons occupy the same shell (*K*).

and so on up to *Q*. If a vacancy should appear in the electron complement of a shell, an electron can, of its own accord, fall into the "hole" (shell vacancy) *only* from a higher energy level (energy flows downhill). For example, a hole in the *K* shell can be filled by spontaneous transition of an *L*-shell electron. On the other hand, energy must be supplied to the atom from the outside to raise an electron from an inner shell to one farther out. When an electron is simply raised to a higher energy level without leaving the atom, the process is called *excitation* (atom in an excited state). When the electron leaves the atom completely, the process is called *ionization* (note that the atom is not only ionized, but excited as well).

The energy or work needed to remove an electron from a particular shell to a point just outside the atom is called the *binding energy* of that shell for the atom in question. Thus, for any given atom, the binding energy is largest for the K shell because it lies closest to the nucleus (recall that the negative electron in the K shell is closer to the positively charged nucleus than is one in an outer shell, so electrostatic force is greater). Each shell in a particular nuclide (atomic species) has a unique binding energy. For example, it is 70,000 eV (70 keV) for the K shell of tungsten, but only 50 eV for the K shell of the average atom in the soft tissues of the body.

Because a relatively large amount of energy, on an atomic scale, is needed to separate an electron from an inner shell, such an electron is said to be tightly bound, that is, a *bound electron*. Conversely, electrons in the outermost shells require virtually no energy for removal and therefore behave as *free electrons*.

Turning to the actual processes by which photons interact with atoms, we may use the following standard classification:

1. Photoelectric Interaction with True Absorption of the Photon.
2. Compton Interaction with Modified Scattering of the Photon.
3. Pair Formation.

We shall now summarize each of these processes.

Photoelectric Interaction with True Absorption

In the photoelectric process (see Figure 3.02) a photon enters an atom and undergoes complete or *true absorption*; in other words, the photon gives up all of its energy to the atom. Immediately, the atom responds by ejecting an electron, usually from the K or L shell, leaving a "hole" in that shell. Now the atom is ionized positively (why?) and it is also in an *excited state*. Note that the energy of the incident (incoming) photon ultimately goes to (1) free an electron from its shell by supplying the binding energy and (2) set in motion the electron as a photoelectron. Thus, the energy of the incident photon is shared (although not necessarily equally) according to the following equation:

$$h\nu_{photon} = W_K + K.E._{photoelectron} \qquad (1)$$

where $h\nu$ is the energy of the incident photon, W_K *is the binding energy of the K shell*, and K.E. is the kinetic energy of the photo-

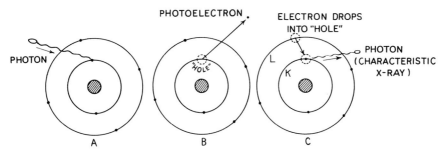

Figure 3.02. Photoelectric interaction. *A*. An incident (incoming) photon gives up all its energy to the atom which ejects a *bound* orbital electron as a photoelectron. *B*. The atom is now excited and ionized since an electron has escaped from the atom. *C*. An electron from some higher energy level fills the "hole" in the *K* shell, with resulting emission of a *characteristic x-ray* photon. (Actually, there is a series or cascade of electrons to fill successive holes, accounting for a series of characteristic photons.)

electron. The holes left in the shells by ejected photoelectrons are filled by transitions of other electrons from outer shells; such transitions are accompanied by the emission of *characteristic* or *fluorescent* x rays whose energy is specific for both the involved atomic species (nuclide) and the difference in the energy levels of the shells between which the electron transitions occur. In other words, a particular kind of atom can emit only a limited variety of characteristic x rays insofar as their energy is concerned. The photoelectrons and characteristic x rays together constitute low-energy *secondary radiation*. Thus, the photoelectrons have an energy of less than 100 keV which, along with their negative charge and their ionizing ability, assures their absorption in only about 1 mm of tissue. (Keep in mind that the photoelectrons contribute to the ionization process in the second step, described below.) The characteristic radiation also has low energy, less than 1 keV, and undergoes local absorption in the tissues. Photoelectric interaction has the greatest probability of occurring between low-energy photons (about 100 to 140 kV beam) and inner-shell electrons (*K* or *L*), in atoms of higher atomic number such as calcium and phosphorus in bone. Note that the probability of a photoelectric interaction is proportional to the cube of the atomic number

(Z^3) of the absorber, and inversely proportional to the cube of the energy ($1/h\nu^3$) of the radiation.

Compton Interaction with Modified Scatter

Here a high-energy photon interacts with an electron in an *outer* shell, much as two billiard balls might collide (see Figure 3.03). The photon *scatters* (goes off in another direction) with less

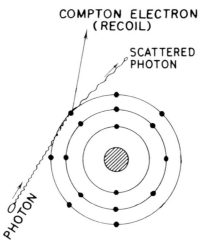

Figure 3.03. Compton interaction with modified scatter of photons. Only a part of the incident photon's energy is used up in imparting kinetic energy to the recoil electron. Therefore, the emerging photon has less energy than the incident photon, and also proceeds in a different direction (scatter).

energy than the incident photon. The electron is ejected as a *recoil electron*. In this case the energy ($h\nu$) of the incident photon is distributed (although not necessarily equally) between the $h\nu$ of the scattered photon and the kinetic energy of the recoil electron:

$$h\nu_{incident\ photon} = h\nu_{scattered\ photon} + K.E._{recoil\ electron} \qquad (2)$$

Only the energy imparted to the recoil electron is truly absorbed, eventually being used up as the recoil electron ionizes atoms in its path. Compton interaction predominates above 50 keV (about 150-kV beam) although it occurs below this level. It is the most important interaction in the megavoltage region, from about 1 to 10 MeV. The probability of a Compton interaction is independent of the atomic number of the absorber.

Pair and Triplet Formation

A photon with a minimum energy of 1.02 MeV may interact with an atom (probably near a nucleus), whereupon it changes into two charged particles: a *negatron* (negative electron) and a *positron* (positive electron) as shown in Figure 3.04. The negatron

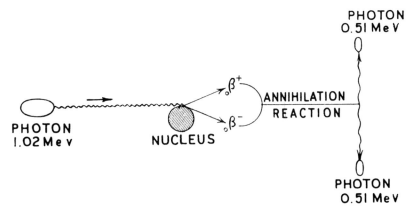

Figure 3.04. Pair production and annihilation process. This takes place near an atomic nucleus. If it takes place near an electron, a triplet is formed consisting of the pair, and the electron that is set in motion during the interaction.

and positron share equally any energy above 1.02 MeV in the form of kinetic energy and go on to ionize other atoms. As the positron comes to rest, upon meeting a negative electron (not necessarily or probably its original mate) both particles disappear. They are immediately replaced by two photons, each having an energy of 0.51 MeV and moving in opposite directions. This process is called the *annihilation reaction*. Note that 0.51 MeV exactly equals the energy of an electron at rest (based on Einstein's equation $E = mc^2$). Pair production becomes an important type of interaction above about 10 MeV.

When a photon with an energy of at least 1.02 MeV interacts

with an electron, a *triplet* is formed. This consists of a negatron and a positron pair, together with the original electron.

Relative Importance of Various Interactions Summarized

When x or gamma rays traverse the body, their interaction with tissue atoms results in the liberation of *primary electrons*. The three main processes include photoelectric interaction with true absorption, Compton interaction with modified scatter of photons, and pair and triplet formation.

Photoelectric interaction predominates below about 150 kV_p. Primary photons in the incident beam undergo complete absorption upon interacting with tissue atoms, setting in motion photoelectrons that ionize and excite tissue atoms along their paths.

As the energy of an x-ray beam is increased from 150 kV_p to 3 MV, and also with ^{60}Co gamma rays (av. energy 1.25 MeV), the probability of Compton interaction decreases, but the probability of photoelectric interaction decreases even more, so the Compton interaction becomes predominant. The resulting recoil electrons go on to ionize and excite tissue atoms.

Although pair and triplet production begins at the threshold energy of 1.02 MeV, only above a photon energy of about 20 MeV does it assume a major role in energy deposition in matter. The resulting electrons cause ionization and excitation of tissue atoms just as do the primary electrons in the other kinds of interactions.

Table 3.01 summarizes the relative importance of the different types of interactions at various beam energies.

TABLE 3.01

SUMMARY OF RELATIVE IMPORTANCE OF VARIOUS INTERACTIONS
WITH DIFFERENT BEAM ENERGIES.

Photon Energy	Approximate Beam Energy	Predominant Interaction
up to 50 keV	180-kV x rays	Photoelectric
50 keV to 8 MeV	180 kV to 10 MV	Compton
24 MeV and up	25 MV	Pair and Triplet Production

STEP 2—Ionization and Excitation by
Primary Electrons

The second step in the deposition of energy by photons takes place when the electrons, liberated by any of the above interactions, move through the tissues. These primary electrons ionize and excite atoms in their path creating *ionization and excitation tracks* (see Figure 3.05). You should recall that ionization implies the complete removal of an orbital electron from an atom, whereas excitation refers to the displacement of an electron to a higher energy level. Obviously, ionization, by removal of an inner shell electron, leaves the atom excited as well. We are uncertain as to the role of excitation in causing biologic effects, and so we customarily speak only of ionization, bearing in mind that excitation also contributes to such effects.

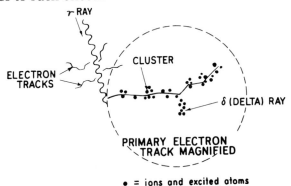

Figure 3.05. Diagram showing an ionization and excitation track of a primary electron that has been liberated by an x- or gamma-ray interaction with matter. Clusters of ions and excited atoms along the primary and delta ray tracks produce the radiobiologic lesions in DNA, the severity of which depends on the closeness of the clusters—the closer they are, the greater the amount of energy deposited per unit length of path.

The density or spacing of ions along a track varies according to the kinetic energy of the electron in question. As the electron's kinetic energy increases, the ionization density *decreases*, strange as it may seem. Conversely, as the kinetic energy of the electron decreases, the ionization density increases. This is explained as follows: as the electron slows down it spends more time in the vicinity of atoms, thereby increasing the probability of ionization. As we shall see later, the effectiveness of a particular kind of radiation in producing biologic effects, that is, the *relative biologic effectiveness (RBE)*, depends on the ion density engendered by the radiation—the greater the ion density (closer spacing of ion clusters) the greater the RBE.

Ionization density along a charged particle track such as that associated with an electron may be expressed in two ways:

1. *Specific Ionization* is the number of ion pairs formed per unit length of path, usually stated as ion pairs per cm.

2. *Linear Energy Transfer (LET)* is the rate of energy deposition per unit of path, usually stated as keV per micron. This is the preferred method because it states in a direct way the distribution of energy deposited along the track and is more closely related to biologic effects. It will receive more detailed discussion in the next section.

PHYSICAL FACTORS IN
RADIOBIOLOGY

The severity of biologic effects engendered by ionizing radiation depends on a number of conditions. These may be broadly divided into biologic, chemical, and physical factors. We shall now discuss only the physical aspects of this subject under four main headings: (1) exposure, (2) absorbed dose, (3) linear energy transfer, and (4) relative biologic effectiveness, all of these being interrelated. The biologic and chemical factors will be described later.

Exposure

As we have seen in Chapter II, exposure is a measure of the amount of ionization produced in a unit mass of air by x or gamma rays and all their associated charged particles. This was discussed in detail on pages 9–10. Although radiation exposure parallels closely the flow of energy in the volume of interest, it does not state the amount of energy actually absorbed.

Absorbed Dose

Because the amount of energy deposited in a given tissue mass strongly influences the intensity of the resulting biologic effects, the concept *absorbed dose* has been introduced to specify such energy absorption.

As we have just pointed out, *indirectly ionizing photon radiation* such as x and gamma rays deposits energy in tissue (or matter in general) in two steps, the second of which involves charged particles (electrons) and is responsible for virtually all the released energy. *Indirectly ionizing particles* such as neutrons act by colliding with atomic nuclei, especially hydrogen nuclei (i.e., protons) which, in turn, cause intense ionization tracks. On the other hand, *directly ionizing charged particle* radiation such as alpha and beta particles deposit energy along their tracks by direct interaction of their electric fields with those of orbital electrons. Recall that alpha particles are positively charged helium nuclei, and beta particles are fast electrons, both these particles being emitted by radioactive nuclei.

The total energy absorbed per 100 grams of matter (i.e., rads) during various interactions, whether by directly or indirectly ionizing radiation, provides the basis for specifying the absorbed dose of x and gamma rays. In general, the absorbed dose governs the magnitude of a particular biologic effect induced by radiation of a particular kind (i.e., value of LET, see below). The rad (or gray) is used to express the energy absorption from any type of ionizing radiation, but as we shall see in the next section, LET is also an important factor insofar as biologic effectiveness is concerned. One rad = 0.01 Gy (gray) in the S.I. system.

Since the absorbed dose (rads or grays) is proportional to exposure (R), except for soft tissue close to bone, we can derive absorbed dose from exposure by using the appropriate roentgen-to-rad conversion factor *f* (obtained from Table 3.02) as follows:

TABLE 3.02

R-to-RAD CONVERSION FACTOR f FOR
VARIOUS GENERATING VOLTAGES
AND RADIOACTIVE SOURCES.*

Energy or Source	Rads per R (f Factor)	
	Muscle	*Bone*
100 kV	0.92	4.14
200 kV	0.94	1.91
250 kV	0.949	1.46
Cesium 137 (0.66 MeV)	0.957	0.92
Cobalt 60 (1.25 MeV)	0.957	0.92
4 MV	0.956	0.92

*Data adapted from Meredith, WJ and Massey, JB. (1977); and from Johns, HE and Cunningham, JR (1978).

$$D = fX \qquad\qquad (3)$$

where D is absorbed dose and X is exposure in R. Note that the f factor applies specifically to conventional units — R and rads. Table 3.03 shows how we derive the factor for converting exposure in R to grays (Gy) in the International System (S.I.). The value of the conversion factor f depends on the average atomic number of the absorber and on the quality of the radiation.

It should be pointed out that equation (3) applies only to the photon energy range up to 3 MeV. Above this limit, the conversion factor C_λ is used to translate the corrected ion chamber values measured in a water phantom to asborbed dose in rads. For 4-MV x rays, the C_λ is 0.94, and for 20-MV x rays, 0.90.

In medical radiology we must understand the difference in the absorbed doses, delivered by a given exposure, as between soft tissue and bone. With low and intermediate energy x rays prevailing in the radiographic and orthovoltage regions, the absorbed dose in compact bone greatly exceeds that in soft tissue, owing to the predominant photoelectric interaction. The probability of a photoelectric interaction is directly proportional to the third power of the atomic number (Z^3) of the absorber, and inversely propor-

TABLE 3.03

CONVERSION FROM EXPOSURE IN ROENTGENS (R) IN
THE CONVENTIONAL SYSTEM, TO ABSORBED DOSE IN GRAYS
(Gy) OR CENTIGRAYS (cGy) IN THE S.I. SYSTEM.

Since	dose in rads = fR
and	dose in Gy = 0.01 × dose in rads
then	dose in Gy = 0.01 fR.
Since	dose in cGy = dose in rads
then	dose in cGy = fR

tional to the third power of the x-ray energy $(1/h\nu)^3$. Since bone contains a large percentage of calcium and phosphorus whose atomic numbers are much higher than the average atomic number of soft tissue, you can see how low-energy x rays give rise to larger absorbed doses in bone than in soft tissue. Thus, with radiographic x rays the f factor in Table 3.02 is four, and using this value in equation (3), we find that the absorbed dose in bone is four times that in soft tissue for the same exposure. In fact, this is the major contributor to contrast in radiography. With orthovoltage x rays (such as HVL 2 mm Cu) the f factor is two, so a 100-R exposure would deliver an absorbed dose of about 200 rads in bone for every 100 rads in soft tissue. On the other hand, with megavoltage radiation, such as cobalt 60 (1.25 MeV) or linear accelerator x rays up to 6 MV, the absorbed dose in bone virtually equals that in soft tissue for equal exposures because in this energy region interaction occurs almost exclusively by the Compton process, which is independent of the atomic number of the absorber.

Linear Energy Transfer

The absorbed dose alone does not tell the whole story. It has been found that equal absorbed doses from different *kinds* of radiation do not necessarily produce the same degree of a stipulated biologic effect. How can we account for this difference in behavior of different kinds of radiation? We know that the distribution, in space and time, of the ions liberated along the path of the radiation governs the intensity of biologic changes (i.e., changes in living matter). The more ions liberated and the more the resultant energy deposited per unit length of path, the greater will be the biologic effect. Table 3.04 shows the large differences in the relative ionizing ability of alpha and beta particles.

TABLE 3.04

COMPARISON OF 2-MeV ALPHA AND BETA PARTICLES; VALUES ARE APPROXIMATE.

Particle	*Average Range in Air*	*Specific Ionization*
	cm	*ion pairs/cm*
alpha	1	60,000
beta	1,000	50

Alpha particles produce much more closely spaced ion clusters along short, straight tracks than do primary electrons or beta particles along zig-zag, longer tracks. Owing to the intense ionization by alpha particles, energy is released over relatively short distances and in only a few cells. In fact, multiple ionizations frequently occur within a single cell, thereby increasing the likelihood of severely damaging or killing the cell (see Figure 3.06).

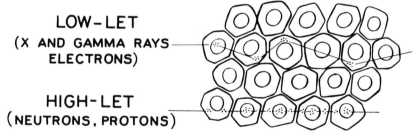

LOW–LET
(X AND GAMMA RAYS
ELECTRONS)

HIGH–LET
(NEUTRONS, PROTONS)

Figure 3.06. Low-LET radiation releases ions and excited atoms (and molecules) cells apart. High-LET radiation releases them in multiple clusters within cells, thereby enhancing the likelihood of inactivating multiple targets and so causing irreversible injury to the cell.

On the other hand, electron tracks consist of ions liberated at relatively long distances from each other, often cells apart, and therefore not only do electrons have a longer range than alpha particles, but multiple ionizations in a single cell are much less likely to occur. Finally, you should recall that photon radiation ionizes indirectly through the primary electrons which they set in motion.

Energy is deposited during the ionization process. The closer the ion clusters, the greater will be the amount of energy released for a given length of path and the resulting biologic effect. This concept is embodied in linear energy transfer, *LET, defined as the quantity of energy deposited per unit length of a charged particle track in matter*. It encompasses both ionization and excitation. Because of the variation in ion density along the track, the density increasing as the speed of the particle decreases, *LET is usually averaged over the entire path of the particle*.

LET has proven to be a very useful concept. We know, for example, that alpha particles produce short, dense, straight ionization tracks associated with a high rate of energy loss per unit length of path; therefore alpha particles are classified as *high-LET radiation*. On the other hand, primary electrons and beta particles have long, irregular ionization tracks with sparse ion clusters, associated with a small energy loss rate; hence, primary electrons and beta particles are classified as *low-LET radiation*.

Some typical average LET values for various kinds of radiation are given in Table 3.05. LET is expressed in *keV per micron* (i.e., per micrometer or 10^{-6} m).

In general, LET increases with an increase in the charge and/or mass of an ionizing particle, but decreases with an increase in the speed of the particle. Because LET is proportional to the square of the particle's *charge*, an *alpha particle* with its double charge would have an LET four times that of an electron of similar speed. Particle *speed* also influences LET; the slower the speed the more time the particle spends in the vicinity of an atom and therefore the greater the probability of interaction and resulting ionization and excitation.

Fast neutrons also represent high-LET radiation because they ionize indirectly by setting in motion recoil nuclei, mainly pro-

TABLE 3.05

AVERAGE VALUES OF LET IN SOFT TISSUE
FOR REPRESENTATIVE TYPES OF RADIATION.

Radiation	Energy	LET
		average keV/µm
gamma rays (^{60}Co)	1.25 MeV (av.)	0.3
x rays	250 kV$_p$	1.5
electrons	1 MeV	0.25
protons	10 MeV	4
alpha particles	5 MeV	100
neutrons	20 MeV	7

tons, on interacting with matter. While protons have the same charge as electrons, they have a mass about 2000 times greater, so that for particles of like energy, the protons move more slowly and are therefore more densely ionizing, that is, protons are high-LET radiation. This is explained by the following equation for the kinetic energy (K.E.) of a particle:

$$K.E. = \tfrac{1}{2}mv^2 \qquad (4)$$

where m is the mass of the particle at rest, and v is its velocity. Assuming that the K.E. of a proton equals that of a particular electron, and knowing that the proton has a much larger mass than an electron, we find the velocity or speed of the proton in our example must be much less than that of the electron for the K.E. of the particles to be the same. Consequently, the much slower proton will have a much higher LET than the electron.

Relative Biologic Effectiveness

Does the difference in LET as between low energy and high energy *photons* produce significantly different biologic effects?

Recall that photons are uncharged and indirectly ionizing. We know, for example, that the average LET of megavoltage radiation with energy of a few MeV is much less than (about 1/5) that of 250-kV x rays, based, of course, on the primary electrons these two types of radiation release. This leads directly to the question of energy dependence—do *photons* of different energies produce biologic effects of different intensities? It is reasonable to suppose that electrons set in motion by low-energy photons would have lower velocity and higher LET than those released by high-energy photons. This has, indeed, been found to be true and has led to the concept of *relative biologic effectiveness* or *relative biologic equivalence*, usually referred to as *RBE*. It applies to all ionizing radiation, whether photon or particle. *RBE is defined as the ratio of the absorbed dose of a standard type of radiation to that of the radiation in question, required to produce the same degree of a stipulated biologic effect.* The standard radiation is customarily 200 or 250-kV x rays.

$$RBE = \frac{absorbed\ dose\ of\ 250\text{-}kV\ x\ rays}{absorbed\ dose\ of\ radiation\ under\ study} \qquad (5)$$

for the same biologic effect.

A particular kind of radiation does not necessarily have a single RBE value; this may depend on such diverse factors as the type of cell or tissue and its physiologic state, the presence of oxygen, the biologic effect under investigation, and the dose rate.

In practice, ^{60}Co gamma rays require a larger exposure for skin erythema (reddening) than kilovoltage x rays: 1000 R for gamma rays, 680 R for 200-kV x rays (HVL 1 mm Cu), and 270 R for 100-kV x rays (HVL 1 mm AL). The generally accepted RBE value for ^{60}Co gamma rays and linac 6-MV x rays is about 0.85. Therefore, in irradiation therapy the absorbed dose in the tumor should be increased about 15 percent with such megavoltage beams to obtain the same effect (e.g., tumor control) as with orthovoltage x rays.

Thus, in medical radiology we must be conversant with a number of physical concepts represented by appropriate units. The roentgen (R) is simply a unit of exposure that can be converted to absorbed dose by the application of a suitable conversion factor (*f*). In the S.I. system, radiation exposure in *exposure units* can be converted to units of absorbed dose Gy by using the conversion factor $2.58 \times 10^{-6} f$.

In summary, then, the fundamental physical process in radiobiology is the transfer of energy to cells during ionization and excitation by interaction of charged particles with atoms. Such energy deposition is expressed in two ways: (1) *absorbed dose*, energy absorbed per unit mass of tissue at the place of interest, and (2) *average linear energy transfer, LET*, the average rate of energy loss (energy deposition) per unit length of path of an ionizing particle. Of the two, the absorbed dose has greater practical importance in radiotherapy at the present time. The LET is valuable in research and in therapy with the newer treatment modalities such as fast neutrons and negative pions since it specifies more adequately the quality of all types of ionizing radiation. As we have indicated previously, the HVL and tube potential in orthovoltage therapy, and the beam energy in megavoltage therapy, satisfactorily specify photon beam quality.

Chapter IV

CELLULAR STRUCTURE AND FUNCTION

As we have seen in the preceding chapter, ionizing radiation releases energy as it passes through matter. In living matter, this excess energy initiates a train of physical and chemical events that injure cells, which subsequently may or may not recover, depending on the severity of the damage, the type of cells, their physiologic (functional) state, and their degree of oxygenation. Thus, the term radiobiology refers to that branch of radiologic science that deals with the mode of action and effects of ionizing radiation on living systems.

To understand better the subject of radiobiology we may approach it on three levels: (1) cellular structure and function, including reproduction, (2) the radiobiologic lesion, and (3) cell and tumor response to ionizing radiation. In this chapter we shall discuss cellular structure and function.

THE CELL

A typical animal or plant cell is a living unit, usually microscopic in size, consisting of two main structural components—a nucleus and its surrounding cytoplasm (see Figure 4.01). While these cell constituents have distinctive characteristics, structurally and functionally, they are mutually interdependent so that any activity or change in one produces an effect on the other.

Nucleus

The nucleus serves as the reproductive center of the cell and regulates metabolism, the totality of life-sustaining cellular func-

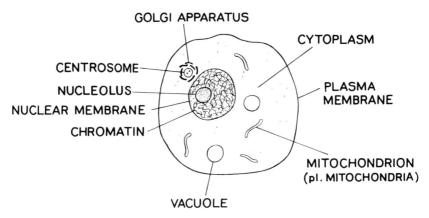

Figure 4.01. Ordinary cell as seen with a *light* microscope.

tions. It is enclosed by a double-layered nuclear membrane which is at least intermittently porous, allowing interchange of essential substances with the cytoplasm. The nucleus contains three important constituents: (1) chromatin, (2) one or more nucleoli, and (3) nuclear "sap." *Chromatin* is a nucleoprotein that stains blue (i.e., basophilic) with basic dyes and consists of an extremely important substance called *deoxyribonucleic acid, DNA*, chemically joined (conjugated) with a simple protein classified as a histone. Chromatin exists as a fine network during cell interphase (i.e., "resting" or intermitotic), but during cell division it condenses to form *chromosomes* which are microscopic threadlike bodies. Thus, the chromosomes consist of a *DNA-protein complex*. They have the capacity to reproduce themselves by a process called *mitosis*.

Each species of animal or plant has a characteristic number of chromosome pairs in its somatic (body) cells. At the same time, each chromosome pair has a somewhat different appearance from the others (see Figure 4.02). Human somatic cell nuclei normally contain exactly 23 pairs of chromosomes, or 46 in all. We may regard a full complement of different human chromosomes as a *set*, so that a somatic cell has two sets. While one set—the *haploid number* designated generally by the letter *n*—is compatible with life, two sets are required for completely normal structure and function of the individual; that is, the *2n* or *diploid number*. The reason for this is the existence of *genes*, which are DNA segments

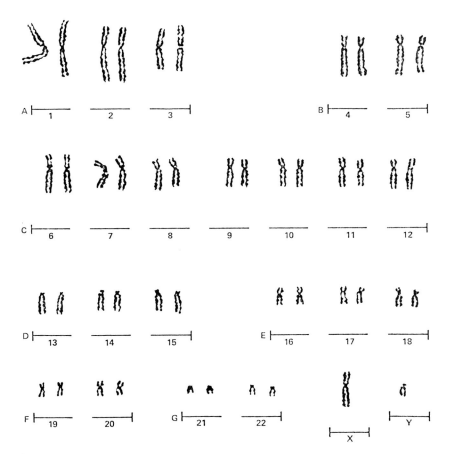

Figure 4.02. Normal *karyotype* (chromosome complement) prepared from human leukocytes dividing in culture. After suitable preparation, the chromosomes appear in a cluster on the slide. A photograph is then made through the microscope and the individual chromosome pictures are cut out with a scissors so the chromosomes can be grouped in homologous pairs. These pairs are then arranged according to size based on the Denver classification. One unpaired chromosome, *Y*, indicates that this is a *male* karyotype. Abnormal karyotypes may have additional or missing chromosomes, aberrant chromosomes, or fragments.

of a chromosome. Genes are lined up like beads on a string and carry the information for the manufacture of various proteins, a subject that will be explored more fully later in this chapter. All somatic cells of a particular species must have the same, characteristic number of chromosomes (i.e., 2n). When by some accident of reproduction or chromosomal aberration there occurs an excess or shortage of chromosomes or parts of chromosomes, relative to the normal number, serious congenital anomalies may result. Changes in chromosomal number or structure influence the quantity of DNA.

Cytoplasm

Surrounding the nucleus is the cytoplasm, which contains microscopic bodies or *organelles* such as the following:

Mitochondria. These microscopic elliptical double membranes are concerned with catabolism (metabolic breakdown) of certain substances, especially carbohydrates, to provide energy for the cell.

Golgi Apparatus. The small, variously shaped bodies consist of double membranes. They are concerned with secretion, carbohydrate synthesis (manufacture), and the bonding of proteins to other organic compounds.

Endoplasmic Reticulum. A double-membrane system of tubes pervades the cytoplasm. As seen under the electron microscope, this exists in two forms: one has a *smooth* surface, while the other is *rough* owing to the presence of ribosomes on its surface. The rough type has to do with protein and enzyme synthesis and other chemical processes, but the function of the smooth type remains incompletely understood, although both types may conduct secretions to the cell surface.

Ribosomes. These bodies are composed of a nucleic acid, *ribonucleic acid (RNA)*, which is intimately concerned with protein synthesis. Some ribosomes adhere to the surface of the rough endoplasmic reticulum, whereas others remain free in the cytoplasm. More will be said later about the ribosomes in the discussion of DNA and RNA function.

Lysosomes. These single-membrane microscopic sacs contain enzymes that assist in the lysis (breakdown or digestion) of substances such as proteins, DNA, and certain carbohydrates. Any condition that increases the permeability of the sac membrane, or causes it to rupture, permits the enzymes to enter the cytoplasm; this may lead to actual digestion of the cell itself, a process called *autolysis.*

Those cells having the capacity to reproduce also contain, in their cytoplasm, a *centrosome* which lies near the nucleus and contains two *centrioles*. The cell as a whole is enclosed in a triple-layered porous membrane through which water and certain dissolved substances can flow into and out of the cell.

CELL REPRODUCTION

Some cell populations continually experience the death of old cells and the reproduction of the young ones of the same kind. Under proper conditions this should go on indefinitely. In 1912 Alexis Carrell started a culture of cells obtained from the heart of a living chick embryo, and maintained it until 1939, at which time it was still thriving. Naturally, excess cells had to be removed from time to time. It is not surprising, in the light of present knowledge about DNA and RNA, and the process of cellular reproduction (mitosis), that the cells at the end of the experiment were remarkably like the original ones.

The mechanism by which somatic (body) cells reproduce has been understood for many years, having first been described by W. Flemming in the early 1880s; in fact, he coined the word *mitosis* to designate this process, centered in the nucleus. However, not much attention was paid to the time intervals in a cell's life *between* periods of mitosis. In 1953, Howard and Pelc suggested a scheme representing a typical cell cycle, shown in Figure 4.03. A typical cell cycle, according to this concept, consists of two stages: *mitosis* and *interphase*. One should not regard interphase as a resting period for the cell, since we know that important processes occur during this interval. Not the least of them is the synthetic phase, called *S*, wherein DNA undergoes replication (i.e., duplication). As shown in Figure 4.03, two gaps occur in the cycle — G_2 before *M*, and G_1 after *M*. Table 4.01 summarizes the active processes that take place during these gaps.

Mitosis

The steps in mitosis (see Figure 4.04) will now be described. Note again that it involves primarily the nucleus, with participation of

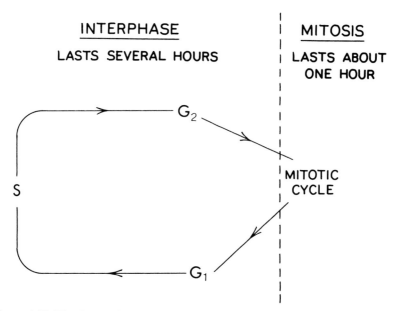

INTERPHASE | MITOSIS

LASTS SEVERAL HOURS | LASTS ABOUT
                       | ONE HOUR

G_2

S | MITOTIC
     | CYCLE

G_1

Figure 4.03. The Howard and Pelc (*Heredity*, 6:261, 1953) concept of a typical cellular generation cycle. There are two main phases—the synthetic *S* phase in which the DNA is normally doubled in amount, and the mitotic *M* phase in which the DNA is normally divided equally between the two daughter cells. The *S* and *M* phases are separated by two gaps, G_1 and G_2, in which other kinds of cellular activity occur, such as RNA and protein synthesis. *S*, G_1, and G_2 constitute the interphase.

the centrosome which lies in the cytoplasm, just outside the nucleus.

Prophase. The DNA, having doubled in amount during the *S* phase, now aggregates in the form of fine, paired threads. These gradually shorten and become tightly coiled and thickened to form the *chromosomes*. Every chromosome pair is joined at some point by a tiny body called a *centromere*, each member of the pair being designated a *chromatid* (see Figure 4.05). Such chromosome pairs are *homologous* since they have the same gene sequence. As already mentioned, the somatic cells of each plant and animal species have a typical number of chromosome pairs, as well as unique chromosomal structure and function. Furthermore, in any one species all the chromosome pairs differ from one another. During prophase the chromosome pairs *duplicate* to form twice the normal diploid number, that is, the 4n number. There are now four

TABLE 4.01

EVENTS DURING VARIOUS STAGES OF THE CELL CYCLE*

Phase	Event
G_1	RNA (all forms) synthesized.
	Enzymes and proteins produced for synthesis of DNA.
	Cells that have lost their ability to reproduce (permanently noncycling), such as nerve and muscle, leave cell cycle *permanently* at G_1.
	Cells that are *temporarily* noncycling leave cell cycle at G_1 and enter G_0.
	Variable duration—length of cycle time varies widely with cell population.
S	Chromosomal DNA synthesis.
	Proteins produced to maintain DNA synthesis.
	RNA synthesis continues at same rate as in G_1 and G_2.
	Maturation and reproduction of centrioles.
G_2	Proteins needed for mitosis are synthesized.
	Synthesis of RNA that will direct synthesis of proteins for mitosis.
	Centrioles divide into pairs.
M	RNA synthesis stops.
	Condensation and segregation of DNA, RNA, and proteins.
	Protein synthesis decreases.
	M phase occupies about 5 to 10 percent of cell cycle.

*Adapted from Dalrymple, GV, et al. *Medical Radiation Biology*, 1973.

chromosomes of each kind called *tetrads*. In humans, the total number of chromosomes at this stage is 92.

While the chromosomes are forming, the nuclear membrane and nucleolus disappear. At the same time, the two *centrioles* lying in the centrosome just outside the nucleus gradually separate and move to the opposite poles of the cell, meanwhile building a *fine spindle* between them. This part of mitosis, up to the instant when the chromosomes have assumed their definitive shape and the centrioles have reached the opposite poles, is called the *prophase*.

MITOSIS

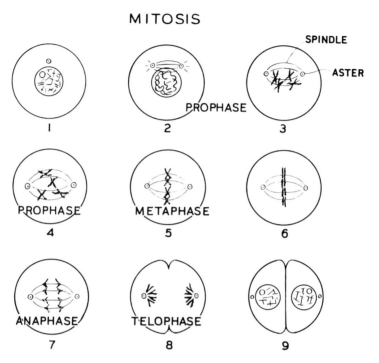

Figure 4.04. Cell division by *mitosis*. There has been a doubling of DNA during the synthetic (*S*) phase, resulting in a doubling of the normal somatic chromosome number to eight in this case (fruitfly *Drosophila*). During mitosis the normal number of somatic chromosomes is restored, that is, four.

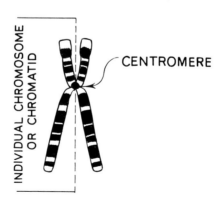

Figure 4.05. Schematic representation of a chromosome pair. The "banding" pattern of the chromosomes, brought out by special stains, increases the accuracy of identifying the various chromosomes.

Metaphase. In the next stage or *metaphase* the chromosome pairs drift toward the equator of the spindle, that is, midway between the centrioles, and arrange themselves in a plane perpendicular to the spindle axis, called the *equatorial* or *metaphase plate.* The chromosomes then become attached to the spindle threads by their respective centromeres. The word mitosis is derived from the Greek word *mitos* = thread.

Anaphase. The next stage or *anaphase* starts with the separation of each chromosome pair into its respective chromatids, each of which moves to the opposite poles of the spindle as the spindle fibers shorten.

Telophase. Finally, during telophase, after the chromosomes have reached their respective pole of the cell, all the chromosomes begin to uncoil and lengthen, reverting to a chromatin network. The nuclear membrane reappears and the cytoplasm divides into two equal portions, each containing a full set (2n) of chromosomes characteristic of the species.

In essence, then, the normal 2n number of chromosomes—46 in humans—doubles during prophase, the amount of DNA having doubled during the *S* phase, thereby providing sufficient material to form a quadruple complement or 4n number of chromosomes. There are now 92 chromosomes in all. During anaphase the chromosome pairs separate and, at random, one of each pair ends up in each daughter cell at telophase. Thus, the original 2n number (46) has been restored, that is, 23 different *pairs* of chromosomes.

The mitotic process lasts about 40 minutes in human as well as in other mammalian cells, two hours in cold-blooded animals, and up to a day or so in plants.

Meiosis

A modified form of mitosis known as *meiosis* governs the reproduction of *gametes* (sex cells, i.e., ova and sperm). During fertilization of an ovum by a sperm these two contribute all their chromosomes to the resulting embryo. If the gametes each contained the somatic (diploid or 2n) number of chromosomes, the embryo

Elements of Radiobiology

would receive a total of 4n chromosomes, or twice the normal number for the species. Moreover, doubling would take place in each subsequent generation, a state of affairs incompatible with survival of the species. Instead, during meiosis the chromatids making up the homologous pairs of chromosomes separate, each going to one of the daughter gametes as shown in Figure 4.06. In

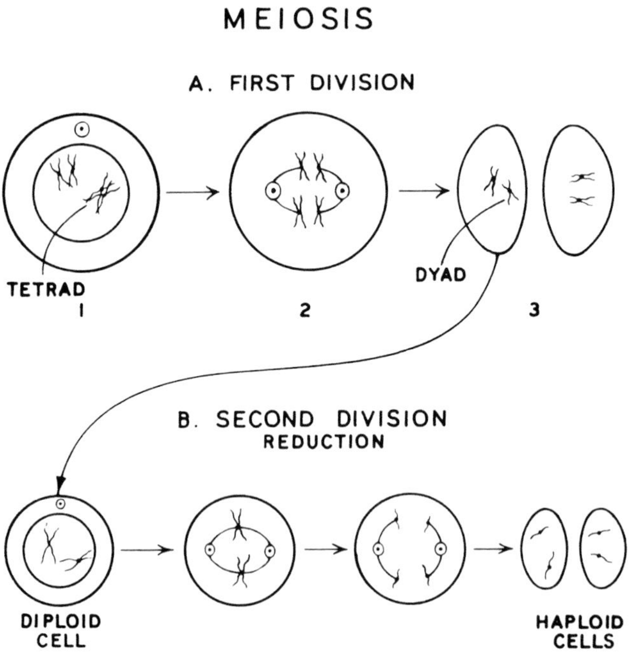

MEIOSIS

A. FIRST DIVISION

B. SECOND DIVISION
REDUCTION

Figure 4.06. Gametogenesis, the formation of reproductive cells (gametes: sperm and ova) by *meiosis*. In this process the normal somatic number of chromosomes (in this case four) is first doubled in the parent cell to form *tetrads*. The first division gives rise to two cells, each containing the normal diploid somatic number (i.e., four chromosomes as in ordinary mitosis). Here the somatic chromosome pairs are called *dyads*. In the second or reduction division the dyads divide so that each daughter cell or gamete contains the *haploid* (half) number of chromosomes.

humans, for example, each mature ovum or sperm contains 23 chromosomes, none being paired—the haploid or n number. During fertilization the ovum and sperm each contribute 23 chromosomes, restoring the 46 or 2n (diploid) number characteristic of human somatic cells (see Figure 4.07). Note that human sperm all contain a *Y* chromosome paired with an *X* chromosome; the *Y* contains few genes. Human ova contain two *X* chromosomes.

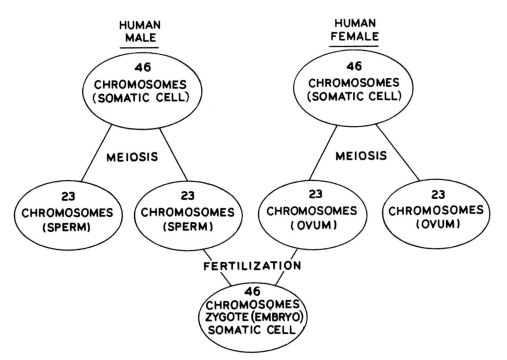

Figure 4.07. Formation of a human zygote (embryo) by fertilization of an ovum by a sperm. During meiosis the gametes (sex cells) have each received one-half the species-specific somatic chromosome number. Union of two gametes, that is, ovum and sperm, during fertilization restores the full complement or diploid number of chromosomes in the somatic cells of the embryo.

MALIGNANT CELLS

We use the general term *cancer* for malignant cells. Such cells have an abnormal structure characterized mainly by an increase in the amount of chromatin, and in the ratio of nuclear material to cytoplasm—*increased nucleocytoplasmic ratio*. However, the overall size of the cell may be unchanged, or it may be larger or smaller than its normal counterpart. In fact, highly malignant cells often display a marked variation in size and shape, a condition called *pleomorphism*. In malignant tumors, cells lose their ability to cohere (stick to each other), favoring their separation and dissemination (spread) to regional or distant parts of the body via the lymphatics and blood stream, and giving rise to secondary tumors called *metastases*. One of the most striking changes observed in cancer cells is an *increase in mitotic activity* as well as the occurrence of abnormal mitoses with unequal division and distribution of chromosomes, and tripolar and multipolar spindles producing bizarre mitotic divisions. Malignant tumors will receive further attention in Chapter XVI.

STRUCTURE OF DNA

The history of DNA (deoxyribonucleic acid) goes back many years. In the 1860s nucleic acid was discovered in cell nuclei, and in the ensuing years the various components of this substance were identified. The acid portion was soon found to be phosphoric acid.

In the early 1900s P.A.T. Levene identified two kinds of sugar in nucleic acid—the pentoses (5-carbon sugars) *ribose* and *deoxyribose*, which characterize two kinds of nucleic acid, *ribonucleic acid (RNA)* and *deoxyribonucleic acid (DNA)*, respectively.

It was not until 1953 that the true physical structure of DNA was clarified. Earlier it had been erroneously thought that DNA consisted of a straight-line (single file) arrangement of subunits called *nucleotides*, each consisting of *phosphoric acid*, a *pentose sugar*, and a *purine* or *pyrimidine*. The phosphate-sugar portion made up the backbone of the molecule, with the nitrogenous bases (purines and pyrimidines) forming a series of sidearms as shown in Figure 4.08. Such a molecule is called a *polynucleotide chain*.

In 1953, an American, J. D. Watson, working with F. H. C. Crick in England, finally elucidated the true structure of DNA (and

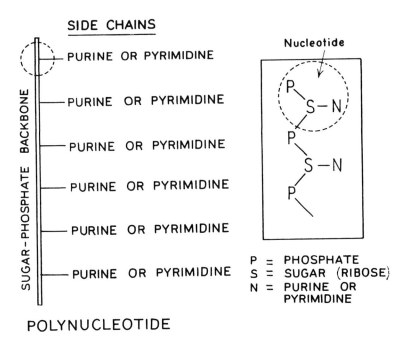

SIDE CHAINS

POLYNUCLEOTIDE

Figure 4.08. On the left is shown a schematic diagram of a straightened single DNA strand consisting of a sugar-phosphate "backbone" and its attached purine and pyrimidine side chains; normally there are two such strands with intervening side chains like rungs on a ladder. The *inset* at the right shows a detail of the DNA structure. (Note that the sugar in DNA is *ribose*, a pentose in which there are five carbon atoms in each molecule.)

RNA). They proved by chemical and x-ray crystallographic studies that the DNA molecule consists of a *double strand* coiled about a common axis, with the sugar-phosphate backbones on the outside and the purines and pyrimidines inside. Known as the Watson-Crick *double helix* (see Figure 4.09), this model of DNA clarified all the hitherto known data that had been accumulated over many years. For example, it accounted for the extremely large molecular size of DNA (molecular weight ranging from about 10^{10} to 10^{12}, or ten billion to 1 trillion). By way of comparison, the molecular weight of oxygen is 16. In DNA fantastic numbers of nucleotides line up in the coiled strands. DNA is therefore called a *macromolecule* (great molecule), consisting of a series of smaller molecules joined

 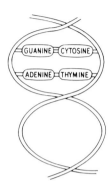

Figure 4.09. Double helix structure (Watson-Crick Model) of a DNA molecule. Horizontal lines represent the weak (hydrogen) bonds between the nitrogenous bases—purines and pyrimidines. This is called *base pairing*.

end to end. Such a molecular combination is called a *polymer* and the joining process *polymerization*. The reverse process, *depolymerization*, refers to the separation of DNA into smaller molecules.

We shall describe now, in as simple a form as possible, the relationship between the two strands in the DNA molecule.

Four Nitrogenous Bases. Only four nitrogenous bases are available for incorporation in DNA: the purines (adenine and guanine) and the pyrimidines (thymine and cytosine).

Pairing of Nitrogenous Bases. The number of adenine and thymine units is always exactly the same, that is, they occur as *complementary pairs*. The same applies also to the guanine and cytosine units. Thus, the purine *adenine* pairs only with the pyrimidine *thymine*, and the purine *guanine* only with the pyrimidine *cytosine*. However, the number of adenine-thymine pairs does not necessarily equal the number of guanine-cytosine pairs.

Parallelism of DNA Strands. X-ray diffraction (scatter) studies of DNA have shown the coiled strands to be absolutely *parallel*, that is, equidistant from each other at each and every point. Purines are longer molecules than pyrimidines, and since there are equal numbers of purines and pyrimidines, the only possible arrangement to maintain strand parallelism is for a purine sidearm of one strand to meet a pyrimidine sidearm from the other strand (see Figure 4.10), each pair being linked by hydrogen bonds. The only possible combinations, as it turns out, are between adenine and thymine with a double bond, and between guanine and cytosine with a triple bond; no other bonding can occur normally.

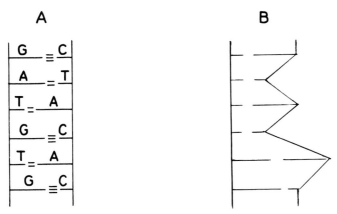

Figure 4.10. In *A* is shown a straightened DNA *double helix*; the two strands are *parallel* and are joined by double and triple bonds between the side chains. The longer *G* (guanine) and the shorter *C* (cytosine) side chains are joined by a triple bond, whereas the longer *A* (adenine) and the shorter *T* (thymine) are joined by a double bond. In *B*, if short side chains were joined, and long side chains were joined, the phosphate backbones would not be parallel and so would not conform to the normal DNA structure.

RNA differs from DNA in having uracil instead of thymine as one of its pyrimidines, and in consisting of a single strand. Furthermore, the RNA strand may have a hairpin configuration with hydrogen bonds between segments of the same strand.

FUNCTION OF DNA

The Watson-Crick model so clearly and correctly explains the structure of DNA that it has been universally accepted and is referred to as *the central dogma*. It also includes the functional attributes of the DNA molecule: (1) replication and (2) coding (instructions) for the synthesis of proteins and enzymes.

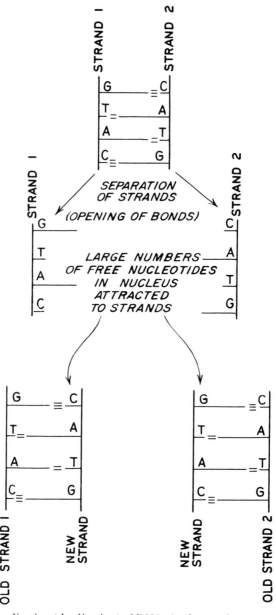

Figure 4.11. Replication (duplication) of DNA. As shown, the two strands separate into individual strands (actually, from one end to the other like a zipper). Each strand then attracts free nucleotides with complementary side chains, *C* to *G*, and *T* to *A*. The new strand is thus complementary to the old, and therefore the new two-stranded DNA molecules have the same nucleotide sequence as the original.

Replication of DNA. During the synthetic (*S*) phase of the cell cycle, the amount of DNA doubles, or *replicates*. In this process, one strand serves as a model (template) for the production of the other strand, the two being complementary, as explained in Figure 4.11A where the double helix has been straightened out for simplicity. Opening of the hydrogen bonds by certain enzymes causes separation of the two DNA strands. The strands do not separate all at once, but rather from one end to the other like a zipper. In the nuclear sap are found large numbers of free nucleotides, which are being continuously manufactured by the cell. Each DNA strand attracts the complementary nitrogenous base-containing nucleotide as shown in Figure 4.11B. Thus, thymine (T) is attracted to adenine (A), cytosine (C) to guanine (G), A to T, and G to C. Hydrogen bonds restore the normal two-strand state in the two daughter DNA molecules which are exact replicas of the original. This is basically what occurs during the synthetic (*S*) phase of the cell cycle. As a result of DNA replication all genetic material (i.e., DNA) should normally be distributed equally to all daughter somatic cells during mitosis. The same applies to early gametogenesis (i.e., early stages of sex cell formation), but in this case only one-half the genetic material reaches the mature gametes during meiosis.

Protein Synthesis. DNA also serves as a *model* or *template* for the manufacture of a great number of specific proteins and enzymes in the cell. Whereas, replication takes place in the nucleus, protein synthesis occurs in the cytoplasm. Furthermore, the latter process requires not only "instructions" from DNA, but also the participation of three kinds of RNA. All cells of a particular animal or plant species contain the same amount of DNA; but the quantity of RNA varies among the cells, being higher in those cells in which there is a higher level of protein synthesis. Thus, growing tissues have more RNA per cell than resting tissues. RNA exists in large concentration in the *nucleolus*, but makes up only about 10 percent of chromosomal nucleic acid.

Although RNA is produced mainly in the nucleus, it diffuses readily through the nuclear membrane into the cytoplasm. There it aggregates with equal numbers of protein molecules present in the form of extremely small particles, about gene size, called *ribosomes*.

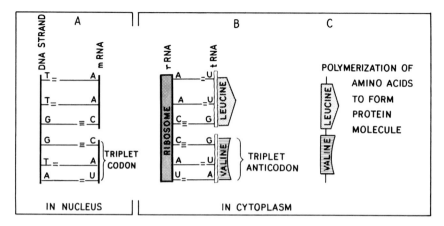

Figure 4.12. DNA as a template for protein synthesis. In *A*, a single DNA strand *in the nucleus* has been joined by two complementary *triplet codons* of messenger RNA (mRNA) according to the rule *T* = *A*, *G* = *C*, but note that uracil (*U*) is present instead of *T* in mRNA.

In *B*, mRNA has moved into the cytoplasm and become attached to a ribosome to form ribosomal or rRNA, which is joined by complementary triplet *anticodons* of transfer RNA (tRNA). The anticodon triplets spell *amino acids* according to the sequence of nitrogenous bases; thus, *UUG* means leucine, and *GUA* means valine.

Finally, in *C*, the amino acids polymerize end-to-end to form a protein molecule.

Figure 4.12 shows diagrammatically the sequence whereby proteins arise by combinations among the 20 naturally occurring *amino acids*, appropriately called the building blocks of protein. These combinations are brought about by the interplay of DNA and the three kinds of RNA. The steps in the process will now be summarized in simplified form.

ENCODING OF MESSENGER RNA BY DNA. In the nucleus one strand of a DNA molecule encodes ("imprints") a type of RNA known as *messenger RNA (mRNA)*. Each mRNA molecule accordingly carries a particular sequence of *three* nucleotides or *triplets* (as directed by DNA) comprising a bit of information called a *codon* (see Figure 14.12A). The sequence (linear order) of the three nitrogenous bases in the codon are complementary to those in the DNA template, that is, C in mRNA corresponds to G in DNA, etc. except for uracil (U) in RNA that is complementary to A in DNA. The mRNA codons join end-to-end to form a long chain of triplets that leaves the nucleus and enters the cytoplasm where it joins a ribosome to form *ribosomal* or *rRNA*.

ASSEMBLY OF AMINO ACIDS BY TRANSFER RNA (TRNA). Now another type of RNA, *transfer RNA*, comes into play. It assembles amino acids to form proteins in accordance with the mRNA codon sequence. Actually there are a number of different kinds of tRNA, each of which will attach itself to one, and only one, particular amino acid present in solution in the cytoplasm. Moreover, the tRNA carrying each kind of amino acid has an attachment site consisting of a triplet sequence of nitrogenous bases, or *anticodon*, which permits this particular tRNA to join rRNA only where the complementary triplet or codon exists. As shown in Figure 4.12B, if the codon triplet on rRNA has the sequence CAU it calls for the tRNA having the anticodon triplet GUA, which specifies that the amino acid valine will join rRNA on the ribosome. In this way, successive triplets bearing the amino acids for which they have been programmed line up in the proper order as originally dictated by DNA.

SEPARATION AND POLYMERIZATION OF AMINO ACIDS. After the amino acids have been assembled, they separate from the nitrogenous bases and polymerize, joining end to end to form the protein molecule (see Figure 4.12C).

The process just described has been likened to speech or writing because it involves a set of instructions by DNA that will eventually produce the required protein or enzyme. For example, the nitrogenous bases in RNA—cystosine, adenine, guanine, and uracil—comprise four letters and from these are formed the codon and anticodon, representing words. With four letters, and three being required for each word, $4 \times 4 \times 4 = 64$ words are possible. Since there are only 20 natural amino acids, and each amino acid represents a word (corresponding to an anticodon), what purpose do the remaining 44 words serve? It is believed that some represent nonverbal instructions such as punctuation, whereas others may simply provide an unused or stored surplus. However, a particular amino acid may be represented by more than one anticodon triplet, that is, by synonyms; but the converse is not true—a particular triplet anticodon can attract only one specific amino acid. Table 4.02 shows examples of amino acids with their corresponding triplet code words. Note, for example, that there are four words for glycine, these being synonyms.

TABLE 4.02

REPRESENTATIVE TRIPLET CODONS ("WORDS")
AND THEIR CORRESPONDING AMINO ACIDS.

Amino Acid	Triplet Codons*
glycine	†GGU, GGC, GGA, GGG
alanine	GCU, GCC, GCA, GCG
tyrosine	‡UAU, UAC
leucine	§UUA, UUG, ‖CUU, CUC, CUA, CUG

*Note synonyms.
†G = guanine. ‡U = uracil. §A = adenine. ‖C = cytosine.

In all this, note that *the specificity of a protein depends on the sequence of nucleotides as programmed initially by DNA*. If even one nucleotide is out of sequence, an entirely different type of protein will be manufactured. Thus, DNA eventually instructs tRNA how to build specific proteins and enzymes. You can see how important this is because proteins constitute a major ingredient, and enzymes perform essential functions, in living matter. In the end, genes— specific segments of the DNA molecule—are responsible for bodily structure and function by way of the models they provide for various kinds of RNA to assemble the appropriate proteins. If a gene should undergo mutation (change), either naturally or artificially, as by bond breakage or change in nucleotide sequence of the DNA, the resulting protein would be expected to be defective and lead to abnormal body structure and function. DNA is extremely vulnerable to bond breakage by ionizing radiation.

Chapter V

MODES OF ACTION OF IONIZING RADIATION ON LIVING MATTER

All biologic systems are damaged by ionizing radiation. Recovery may occur if such radiation is given in amounts small enough to cause minimal damage. However, beyond this dosage level, radiation can produce biologic changes from which recovery is impossible. In the intact animal this may result in serious aftereffects such as decreased life span and increased likelihood of cancer. In addition, injury to gametes may lead to genetic defects in later generations.

As we have already seen, x and gamma rays release primary electrons in tissues, and these electrons, in turn, cause ionization and excitation of atoms as well as breakage of molecular bonds along their paths. Thus, energetic *photon* radiation ionizes *indirectly* in that the primary electrons which they release actually become the principal ionizing agents.

We do know that a very small amount of ionizing radiation may cause detectable injury to cells. To put this in perspective we should realize that a total human body dose of only 500 rads increases body temperature by only 0.001 C; yet this dose is highly fatal, causing death in over 50 percent of exposed individuals.

H. Kaplan (1972) has described killing by radiation as a chain of amplified reactions:

1. Minute (extremely small) amount of radiation energy triggers, in about 10^{-6} (or one one-millionth) second—
2. Chemical changes in certain macromolecules (very large molecules) such as DNA-protein, and in water, causing—
3. Radiobiologic lesions which may be modified by—
4. Repair processes, not all of which are specific for radiation injury.

55

The mechanisms by which radiation causes the *radiobiologic lesion*, that is, the radiation-induced changes in the DNA, can be understood more readily by analogy with the effects of radiation on a solution. Both cells and solutions exemplify *multimolecular systems*. A simple solution consists of a substance called a *solute* dissolved in another substance called a *solvent*. An example would be a solution of sugar (solute) in water (solvent). Of course, a solution may contain multiple solutes in the same solution, hence the term *multimolecular system*.

When ionizing radiation passes through a solution, ionization occurs in a random (hit-or-miss) manner so that those molecules present in larger number will have a greater chance of being "hit." For example, if the number of molecules of a substance A exceeds that of substance B, then more molecules of A than B will be struck by ionizing radiation. Since the number of water molecules usually exceeds that of the solute, especially in a dilute solution, you can see that there is a greater likelihood of water molecules being ionized than of solute molecules. At the same time, the total number of molecules ionized by radiation depends directly on the absorbed dose, up to a dose that is large enough to ionize them all. So at any dose below the maximum there will be molecules that remain un-ionized.

However, solute molecules not receiving direct hit by radiation do not necessarily escape damage, for energy can be transferred to them by ionization products of other nearby molecules such as water. This occurs as follows: ionization of water gives rise to extremely reactive products called *free radicals*, which can induce profound changes in nearby solute molecules. The term *radiolysis of water* encompasses the various changes induced in water by ionizing radiation. (This subject will be pursued in greater detail below.)

Cells consist mainly of water, which range in content from about 70 percent in skin to 85 percent in muscle. Thus, water is the solvent in a cell. The other constituents, including proteins, enzymes, carbohydrates, lipids, nucleic acids (DNA and RNA), etc. are either dissolved (solutes) or in some way dispersed in the water. These solutes may undergo ionization directly by the radiation (e.g., primary electrons, alpha particles), or indirectly by the free radicals released through the radiolysis of water. The *indirect process predominates* owing to the large excess of water relative to solute in cells.

The solutes in living cells differ in their relative importance for cell integrity and survival. For example, enzymes (special kinds of proteins that speed up specific chemical reactions) are usually present in great abundance so that damage to them ordinarily has a minor impact on cell function. On the other hand, DNA occurs in such limited quantity that damage to one or a few DNA molecules may produce disproportionately large effects on the cell. For example, radiation injury to DNA may cause a much greater loss of enzymes (through impaired production) than by direct irradiation of enzymes.

MODES OF ACTION

We shall now discuss the distinct categories of direct and indirect modes of action of radiation on living systems, but it must be understood at the outset that both modes of action exert their effects on certain vital "targets" within the cell. Only the mechanisms are different. Still, from the historical standpoint direct action has often been termed the *target theory*.

Direct Action ("Target" Theory)

This presupposes the existence of highly radiosensitive (i.e., responding readily) foci or targets somewhere in the cell. At present, most evidence indicates that more than one target must be hit to inactivate the cell by loss of reproductive activity or actual cell death. We designate this as a *multitarget model*. The generally accepted target material is *deoxyribonucleic acid (DNA)* in the cell nucleus. As a result of direct action we may find injury to gene DNA causing a *mutation*, which may or may not cause cell death (lethal effect). Or, the ionization track may induce chromosome breaks with resulting abnormal configurations called *aberrations*, or atypical distribution of chromosomes during mitosis. In general, direct action is more likely to occur with heavy or highly charged particles (for example, neutrons, protons, alpha particles) on account of their higher LET with close spacing of ion clusters. This improves the chance of hitting multiple targets and inactivating the cell. In other words, we have a bird-shot effect.

Indirect Action

As already mentioned, important changes occur in the intra-cellular water (i.e., water within cell) during the passage of ioniz-ing radiation. The induction of such changes is called *radiolysis of water*. Various cytotoxic (i.e., poisonous to cells) substances are released during radiolysis, which lasts about 10^{-12} (one million-millionth) second, the most important being the *free radicals* H· and OH·, and the *solvated (hydrated) electron* e^-_{aq}. These are produced in the following sequence:

a. Ionization by removal of orbital electrons.

$$H_2O \xrightarrow[energy]{radiation} H_2O^+ + e^-$$
$$\text{water} \qquad\qquad \text{water ion} \quad \text{electron}$$

b. Liberated electron e^- immediately becomes surrounded by water molecules to form solvated electron.

$$e^- + 4H_2O \rightarrow e^-_{aq}$$
$$\text{solvated electron}$$

c. Positive water ion H_2O^+ from step (a) promptly dissociates.

$$H_2O^+ \rightarrow H^+ + OH\cdot$$
$$\text{hydrogen} \qquad \text{free}$$
$$\text{ion} \qquad\qquad \text{hydroxyl}$$
$$\text{radical}$$

d. Electron joins a water molecule, forming a negative water ion.

$$H_2O + e^- \rightarrow H_2O^-$$

e. Negative water ion dissociates.

$$H_2O^- \rightarrow OH^- + H\cdot$$
$$\text{ordinary} \qquad \text{free}$$
$$\text{hydroxyl} \qquad \text{hydrogen}$$
$$\text{ion} \qquad\qquad \text{radical}$$

As a result of the indirect action of radiation, the extremely reactive free radicals and solvated electrons transfer energy to molecules that, like DNA, are essential cellular components. Radiation injury is produced largely by the breakage of chemical bonds in such molecules. The free radicals and solvated electrons have a very short life — about 0.1 to 1 microsecond in tissue, and up to 100 microseconds in pure water. Because of their ability to diffuse (spread) they can exert their effects at short distances from their points of origin, in contrast to the mechanism of direct action in which energy is deposited by ions directly in the target macromolecule DNA.

Owing to their marked reactivity, free radicals combine not only with other molecules, but also among themselves to form other radicals or to produce hydrogen peroxide, itself quite toxic to cells. *Oxygen* enhances free radical injury to cells, thereby improving the response of tumors to photon radiation. Of major importance is the ability of free radicals to attack sensitive cellular targets such as DNA and enzymes. Thus, even though chromosomes and DNA may suffer profound damage by ionization accompanying direct action of radiation, injury occurs with much greater frequency by free radicals generated during indirect action of radiation.

In summary, the indirect action process involves transfer of energy to water molecules with prompt formation of free radicals and solvated electrons. These highly reactive entities, in turn, decompose the target macromolecules — DNA — mainly by breakage of bonds, entailing cellular injury. Figure 5.01 shows the mechanism of direct and indirect action.

REACTIONS OF FREE RADICALS

The following equations exemplify some of the interactions of free radicals for those who may be interested in this aspect of radiobiology.

1. *Reactions Among Free Radicals*

$$H \cdot \; + \; H \cdot \; \rightarrow H_2 \qquad\qquad \textit{(molecular hydrogen)}$$

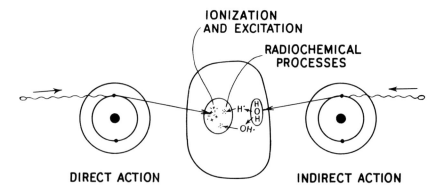

Figure 5.01. Mode of action of radiation on a cell. By direct action (shown on the left) a charged particle enters the nucleus and ionizes and excites the sensitive target (DNA). On the other hand, indirect action (shown on the right) involves the interaction of charged particles with water to produce free radicals such as H· and OH· , and solvated electrons; all of these can injure a sensitive target. Actually, these have a short range so if they are to reach DNA they must be released within the cell nucleus.

$$H· + OH· \rightarrow H_2O \qquad \text{(water)}$$
$$OH· + OH· \rightarrow H_2O_2 \qquad \text{(hydrogen peroxide)}$$

2. *Reactions Between Free Radicals (F·) and Oxygen*

$$F· + O_2 \rightarrow FO_2^- \qquad \text{(superoxide or hyperoxal radical)}$$

To carry this reaction farther,

$$FO_2^- + H· \rightarrow F_2O_2 \qquad \text{(peroxide radical)}$$

3. *Reactions Between Free Radicals and Organic* molecules. The symbol for an organic molecule is $R-H$ where $(-)$ represents hydrogen bonding.

$$R-H + OH· \rightarrow R· + H_2O$$
$$R-H + H· \rightarrow R· + H_2$$
$$R-H + HO_2^- \rightarrow R· + H_2O_2$$

R−H can also change to two free radicals by indirect action of ionizing radiation in which R−H → R· + H· . If R−H should happen to be DNA, essential as it is to the chemical life processes in the cell, direct and indirect action of radiation should cause severe disturbances in cellular metabolism and reproduction.

4. *Reactions Between Organic Free Radicals and Oxygen.* As will be seen later, oxygen enhances the response of cells to low-LET (photon) radiation. This probably involves not only the interaction between inorganic free radicals and oxygen (see under 2) but also a similar interaction involving organic free radicals:

$$R-H \xrightarrow[\text{radiation}]{\text{ionizing}} R\cdot + H\cdot$$

Then R· + O_2 → RO$_2$, an inactive free radical which prevents "repair" of R· that would otherwise occur in the absence of oxygen as follows:

$$R\cdot + H\cdot \rightarrow R-H \text{ (normal molecule)}$$

Note that the *oxygen effect*, as just described, is most pronounced with the indirect mode of action of radiation.

Because of the delicate balance among the numerous cellular components developed over millions of years, a change in certain ones may induce profound changes in cell function, structure, or reproduction. This holds true especially for substances present in small supply and that happen to be essential for cell survival, such as DNA. Hence, no amount of ionizing radiation may be considered "safe." Even an extremely small dose has the potential of causing irreversible damage to an essential cell constituent, although the net effect on the total organism depends on the importance of that particular cell for the body as a whole.

Chapter VI

RESPONSE OF CELLS TO IRRADIATION

When heavily irradiated cells are observed under a microscope, they may be found to have undergone one of two kinds of *necrosis* (death) *not unique* to irradiation. Cells undergoing *coagulation necrosis* stain poorly with the usual dyes, the nucleus breaks down, and the cytoplasm loses its usual fine structure. These changes become visible in a few hours and reach a peak in a few days. The other type of cell death occurs typically in lymphocytes within follicles and is called *liquefaction necrosis*. Observable within a few minutes after irradiation, the changes consist of pyknosis (condensation of chromatin), karyorrhexis (nuclear fragmentation), and change in the cytoplasm from a granular form to a homogeneous liquid. In less sensitive cells these effects are modified in that the cell swells and the chromatin becomes dispersed in the cytoplasm.

On the other hand, there are certain changes that result *specifically* from exposure to ionizing radiation. Before describing them, we should emphasize that radiation is *always injurious to cells*, the severity of the insult depending on the absorbed dose, LET, cell radiosensitivity, cell oxygenation, and position of the cell in the reproductive cycle. These factors will receive a great deal of attention later. For the present, we shall turn to the primary radiation effects on the nucleus itself; radiation changes in the cytoplasm are ordinarily secondary to those in the nucleus.

CELL NUCLEUS—RESPONSE TO IRRADIATION

Because the nucleus acts as the cell's control center, we should anticipate that nuclear radiation injury would have profound effects on the cell as a whole. This turns out to be true, the radiation dose required for severe nuclear injury being much less than that for the cytoplasm. Nuclear radiation injury entails the following changes: (1) mutations, (2) delay or inhibition of mitosis, (3) abnormal mitosis, (4) formation of giant cells, and (5) prompt lysis.

Mutations

A mutation may be defined as a sudden heritable change in a chromosome or gene, resulting in a change in some trait that is transmitted to offspring. Mutations do not always become apparent immediately, but may require several generations before they express themselves. Mutations are of two types: (a) *chromosome mutations*, preferably called *aberrations*, characterized by a change in the chromosomal number or physical structure, and (b) *gene mutations* consisting of chemical changes in the DNA molecule itself. Mutations may be *spontaneous* or *induced*, the latter being especially associated with ionizing radiation, although they can also result from exposure to a variety of chemicals.

Chromosome Aberrations

One of the main radiation effects is the chromosome *break* which, as the term implies, involves an actual separation of a chromosome into two or more fragments (see Figure 6.01). This results in unequal distribution of chromosomal material (i.e., DNA) to the daughter cells during subsequent mitosis. In other words, one daughter cell receives too much DNA while the other

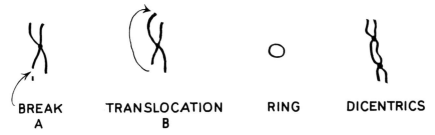

BREAK   TRANSLOCATION   RING   DICENTRICS
A   B

Figure 6.01. Several types of chromosome aberrations. In *A* is shown a chromosome fragment resulting from a chromosome "break" by radiation. In *B* the fragment has become attached to another chromosome, a process called translocation. In the ring form, a single chromosome joins at its ends. A dicentric results from fusion of two chromosomes end-to-end.

one receives too little, leading to various degrees of impairment of cell structure, function, and survival. In some instances, chromosome breaks, especially when occurring in a single strand, may undergo spontaneous repair, thereby restoring the normal chromosomal state. Other aberrations include *stickiness* and fusion of the whole chromosomes or fragments, again causing unequal distribution to daughter cells. In addition, *interchange* of fragments may occur between neighboring chromosomes. Figure 6.01 shows just a few of the more common chromosomal aberrations induced by ionizing radiation. Keep in mind that chromosome aberrations result in a change in the amount of DNA in the daughter cell nuclei.

Gene Mutations

The existence of genes was suspected many years ago as the result of G. Mendel's breeding experiments with sweet peas (1865), and T. H. Morgan's with the fruitfly *Drosophila* (1910). Such experiments revealed that specific points along each chromosome governed corresponding bodily traits or *phenotypes*. The name *gene* was applied to any specific chromosomal site or locus. Each particular gene has a matching gene on the other chromosome of each homologous chromosomal pair, such matching genes being called *alleles* (*alels*) as in Figure 6.02. Now, this does not mean that each gene of an allelic pair has the identical trait-producing effect. Actually, one gene may specify the trait blue eyes, while its mate specifies black eyes; the individual bearing these particular genes will have black eyes, because the trait black eyes is *dominant* over blue eyes. Conversely, the trait blue eyes in this example is *recessive* to black eyes. On the other hand, in some instances incomplete or intermediate dominance occurs, as in a type of chicken known as the Blue Andalusian Fowl. These birds have blue feathers, but their color is determined by an allelic gene pair, one for white and the other for black feathers. Recall that the allelic genes (as part of the chromosomes) are derived from the parents, one from the mother and the other from the father.

It must be emphasized that allelic genes do not exist only as two kinds. Changes in the chemical structure of a gene may give rise to multiple alleles. However, a nucleus normally contains

Figure 6.02. Example of alleles: *a* and *b* are genes in identical loci (positions) on homologous chromosomes (those that pair during mitosis). The genes thereby control the corresponding physical or mental traits in the individual; such manifest traits are called *phenotypes*. Paired genes such as *a* and *b* are called *alleles*. If gene *a* dictated "blue eyes" and gene *b* "brown eyes" the individual would have brown eyes because the gene for brown eyes is dominant.

only two alleles corresponding to the homologous chromosome pair to which they belong.

Figure 6.03 explains the transmission of genes from parents to offspring, when one gene happens to be dominant. We have used a simple example, namely, the traits *normal wings* and *vestigial* (very short) *wings* in the fruitfly (*Drosophila*). In this particular normal-winged parent both genes for normal wings are present (one on each homologous chromosome), so the individual is *homozygous* for normal wings. The other parent, having short wings, must be necessarily homozygous for the recessive short wing trait, having both genes for short wings. Note that the first set of offspring, or F_1 *generation*, all have normal wings because in each individual the allelic genes for normal wings and short wings are present and the normal-wing gene is dominant. In other words, these flies are *heterozygous* for this trait. If two of these F_1 flies are mated, their offspring (F_2 generation) will have the genetic makeup shown in Figure 6.03; now the ratio of normal-winged to short-winged flies is 3:1. For example, on the basis of statistical chance, because of the random distribution of the chromosomes bearing the genes, if there were 24 offspring of the F_1 mating, 18 would have normal wings and 6 would have short wings under ideal conditions. Actually, the 3:1 ratio would be approached only if there were an extremely large number of offspring because of the way the laws of chance operate.

This mode of transmission of traits from parents to offspring is called *Mendelian inheritance*, after the Austrian monk Gregor Mendel

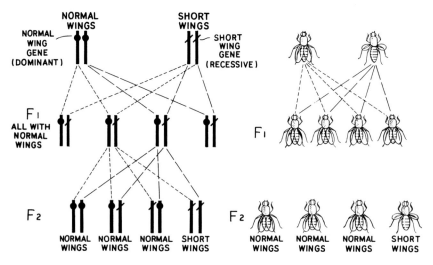

Figure 6.03. Mendelian inheritance. On the left is shown schematically the inheritance of normal and short wings in the fruitfly (*Drosophila*). A normal-winged parent carrying the dominant gene for normal wings on each of the paired chromosomes (homologous for normal wings) is crossed with a short-winged parent, which carries the recessive gene for short (vestigial) wings on the two homologous chromosomes.

The first (F$_1$) generation flies all have normal wings—they are heterozygous and carry the dominant gene for normal wings. The distribution of the corresponding genes is shown by long and short dashed lines.

Mating of two F$_1$ flies will produce, in the next or F$_2$ generation, an ideal ratio of three normal flies to one short-winged fly.

On the right are shown the flies corresponding to the schematic stick figures on the left.

and his experiments with sweet peas. It is truly remarkable that long before the discovery of DNA and the genes he had derived the principles of genetic inheritance.

With the discovery of the true composition and nature of DNA, gene structure and function were immediately clarified. *A gene is simply a DNA segment having a particular sequence of nucleotide bases*, that is, purines and pyrimidines (see pages 46–49). Genes are lined up along chromosomes like beads on a string, their DNA composition making them especially liable to radiation injury, manifested by a mutation. The altered gene is a *mutant*, and the process by which mutation occurred is termed *mutagenesis*.

Radiation mutagenesis was discovered by H. J. Muller in 1927 at the University of Texas in experimental x-irradiation of fruitflies. He found that the type of mutations produced in this way resembled those arising spontaneously in nature; radiation merely increased the frequency of mutation and not the kind. Muller received the Nobel prize for his monumental discovery.

Based on present knowledge about the nature of DNA, and the mechanisms by which ionizing radiation deposits energy in living matter (see pages 55–59), the following effects have been suggested to explain radiation mutagenesis, although this process is not yet completely understood.

Base Deletion (Loss). Radiation may completely remove one or more bases from the DNA molecule. This may lead to *frame shift* in which successive triplets in DNA move over to fill the gap left by the missing base as shown in Figure 6.04. Information lost in this manner has an effect similar to the writing of a meaningless or garbled sentence. Thus, genetic information could be hopelessly confused, unless the missing base has only a minor role to play in the genetic code.

Base Substitution. One base may be substituted for another in DNA. For example, instead of adenine pairing with thymine (the normal situation), adenine may pair with cytosine (normally "forbidden"). Not only is such altered DNA permanent, but it is transmitted to succeeding cell generations. This type of mutation occurs during the DNA synthetic (*S*) phase.

Base Change. Radiation may alter the chemical structure of a base either by direct ionization or by free radicals. The changed base obviously forms "strange" DNA with resulting incorrect encoding of RNA.

As we have stated before, mutations, regardless of how they are produced, usually harm the individual, or if in the gametes, they harm the offspring. This comes about through loss of information or the appearance of incorrect information in the genetic code resulting from changes in DNA such as those just described. One manifestation of mutation is the *impaired viability* of the cell harboring the mutant gene or chromosome. The term viability, in this sense, refers to the physical and functional fitness of the cell or individual and the ability to recover from injury or disease. Depending on the number and importance of the mutations, the effect on cell viability may range from some degree of life shorten-

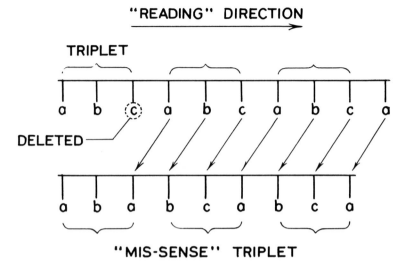

Figure 6.04. *Frame shift.* In the upper half of the diagram we see a single strand of DNA with a sequence of triplets or "words." If one nitrogenous base, *C*, is deleted new triplets are formed as shown, thereby changing the sequence. This changes the character of the DNA which now gives "mis-sense" or nonsense instructions for the production of proteins.

ing to prompt death (lethal mutation). A lethal mutation obviously disappears because it cannot be transmitted to the next generation. However, other mutations cause varying degrees of disability and may pass on to succeeding generations. Eventually such mutations disappear, the more detrimental ones doing so before the milder ones, by a gradual "weeding out" process.

The term *viability* as applied to *cell survival* following experimental irradiation connotes the ability of such cells to continue reproducing. Loss of viability in this sense refers to the survival of the *cell line* rather than to the individual cell, so it has a far different implication than lethal mutation which applies to the cell itself.

Medical and dental x-ray examinations increase the gene mutation rate, that is, the number of mutations occurring per generation. In fact, the mutation frequency increases in proportion to the radiation dose, so that doubling the dose will eventually double the mutation rate of a particular gene. However, the overall incidence of such mutations in diagnostic radiology must be extremely small. *Somatic mutations*—those in the individual's body cells—give rise to abnormal cells and tissues, and may induce *cancer*, if the mutation arises in cells capable of reproducing. Although cancer induction rarely occurs with doses prevailing in radiography, this possibility should not be ignored. In any event, the effect occurs in the individual. However, *mutations in gametes* (sex cells) are transmitted to future generations and if the number of mutant genes becomes large enough, harm could result to the species. This is the main reason for striving to limit exposure, particularly to the gonads, during diagnostic medical and dental x-ray examinations. (See also pages 211–212.)

There is at present considerable evidence for a dose-rate or dose-fractionation effect on mutation frequency, but in practice we assume there is no such effect, to err on the side of safety. Repair of DNA does occur at extremely low dose rates, especially when only one strand has been injured.

Again we must emphasize that radiation effects are *cumulative* insofar as gene mutations are concerned, so that the total accumulated dose determines the mutation rate in a particular gene. Finally, there is no threshold below which mutations do not occur—any dose, no matter how small, has mutagenic potential. (See also pages 184–186.)

Delay and Inhibition of Mitosis. Radiation causes delay in the onset of mitosis if given during interphase, especially in the synthetic (S) phase. However, cells already in the later stages of mitosis may continue through the mitotic cycle. Radiation may completely inhibit mitosis so that a cell permanently loses its ability to reproduce. Figures 6.05 and 6.06 show the effect of radiation on mitosis in a population of proliferating cells. Finally, damage to DNA may not show up until the next mitosis, during which the cell may die—so-called mitotic death.

Abnormal Mitosis. This may be exemplified by *anaphase lag* in which one chromosome of a homologous pair *lags* behind on the mitotic spindle during anaphase. As a result, one daughter cell receives both of these chromosomes, while the other daughter cell ends up without this particular chromosome. Another abnormality, *nondisjunction*, produces the same effect due to the failure of the chromosome pair to separate during metaphase.

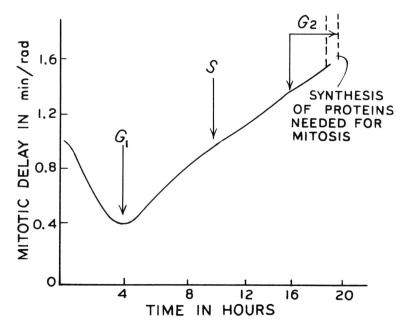

Figure 6.05. *Synchronized* cell population (all cells in same phase of cell cycle), response to irradiation. Exposure to x rays at four hours after synchronization (cells now in G_1) causes the least mitotic delay. Irradiation at ten hours (cells in S phase) causes intermediate delay in mitosis. At 16 to 20 hours irradiation entails the greatest degree of mitotic delay if applied early, before the proteins needed for mitosis have been manufactured—this is the so-called G_2 block. Note that the G_2 period is very short. (Adapted from Terasima T, Tolmach LJ. *Biophysical J* 3:11, 1963.)

Giant Cell Formation. Some cells, after irradiation, may continue to synthesize DNA but do not undergo mitosis. Thus, they become nonreproducing giant cells, which can no longer contribute to the growth of the cell population.

Interphase Death. Most radiosensitive cells die during the next mitosis following clinical doses of radiation. *Mature lymphocytes* are an exception to this rule: while they are rarely seen to undergo mitosis, they die within hours after small doses of radiation. Their death must therefore occur during interphase, and so this lethal response is called *interphase death* or *prompt lysis*. Rossi has suggested that the entire lymphocyte serves as a target, rather than

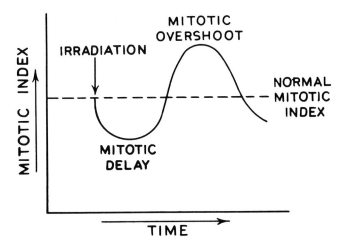

Figure 6.06. *Asynchronous* cell population, response to irradiation. A cell population is ordinarily asynchronous, cells being in various phases of the cell cycle, and the mitotic index (fraction of cells in mitosis) is nearly constant. Irradiation causes maximum inhibition of cells in interphase and early prophase (i.e., maximum mitotic delay). As these cells recover, they enter mitosis along with the unaffected cells leading to an excess frequency of mitosis called *mitotic overshoot*. Finally, the cell population reverts to its normal mitotic index. Thus, there has been temporary synchronization of mitosis by irradiation. (Adapted from Pizzarello FG, Witcofski RL [1975] based on Canti RG, Spear FG [1929] with chick fibroblasts.)

the nucleus alone. It should be pointed out that not all lymphocytes in a given population show this type of response; *in vitro*, about one-half the lymphocytes exposed to 100 R undergo interphase death.

In summary, then, we may say that ionizing radiation induces changes in the DNA macromolecule leading to distorted information for the production of proteins and enzymes that are essential for cellular structure and function. Such abnormalities in DNA constitute gene mutations. Another type of radiation injury to genetic material is manifested by chromosome aberrations such as breaks, interchanges, stickiness, etc. Other effects include delayed mitosis and mitotic death, abnormal mitosis, giant cell formation, and interphase death.

CYTOPLASM—RESPONSE TO IRRADIATION

Although we have emphasized that ionizing radiation induces changes predominantly in the nucleus, manifested by damage to DNA, the cytoplasm is not entirely spared. Cytoplasmic injury may be structural, functional, or both.

The *cell membrane* may show a change in *permeability* even after small doses of radiation, so that the exchange of important substances between the interior of the cell and its environment is impaired. With increased permeability of the cell membrane, especially with larger doses, excess water enters the cell until it is lysed (broken down).

Certain important structures in the cytoplasm evince injury with doses of several thousand rads. For example, the *mitochondria* may be so altered by swelling and membrane damage that their metabolic function is curtailed. Similarly, the *endoplasmic reticulum*, a collection of membranes, undergoes swelling. Injury to the *Golgi apparatus* may interfere with the transfer of secretions to the cell surface.

Note that the harmful effects on the cytoplasm are not specifically manifested in a target—as DNA in the nucleus—but rather in a general way in various structures. Common to the various aspects of cytoplasmic damage is swelling and altered membrane permeability. Since their radiosensitivity is so much less than that of the nucleus and since doses as large as 25,000 rads to the cytoplasm with the alpha particles do not impair the reproductive capacity of cells (Hall), cytoplasmic effects of radiation are ignored in practice.

Chapter VII

RELATIVE RADIOSENSITIVITY
OF CELL POPULATIONS

Experience with radiotherapy has shown that tumors vary in radiosensitivity; that is, some tumors disappear more rapidly or more completely than others when exposed to the same dose of ionizing radiation. Similarly, normal tissues and organs vary in their susceptibility to radiation damage.

How can we account for such a difference in behavior of various tissues? Much of the answer lies in the reproductive behavior of cell populations, which may be divided into two components: (1) *cell cycle kinetics*, or the movement or nonmovement of individual cells around the cell cycle, and (2) *cell population kinetics*, or the movement or nonmovement of primitive cells to more mature types.

Cell Cycle Kinetics

In Figure 4.03 on page 40 we introduced the classical model of the cell cycle as first proposed by Howard and Pelc (1953). A modified version has appeared in recent years, giving a more realistic picture (see Baserga, 1981). According to this revised model, cell populations consist of three main types, exemplified by active bone marrow.

Cycling Cells. These are actively proliferating (reproducing) cells that progress from one mitosis to the next, through successive phases of the cell cycle, to the next mitosis. The phases include mitosis, G_1, S, G_2, mitosis, etc. One cell cycle lasts from the midpoint of one mitosis to the midpoint of the next mitosis. In active bone marrow, the normally cycling cells are the *intermediate* dividing and differentiating cells (see page 76). The cells of most malignant tumors also cycle in this manner.

Temporarily Noncycling Cells. These are represented by *stem cells* in the bone marrow, most of which do not cycle but remain quiescent in the G_0 phase until a demand is placed on them. Thus, when the mature circulating white blood cells, specifically the granulocytes, decrease below a critical number required for body needs (as in combating bacterial infection, or following bone marrow depression induced by ionizing radiation or certain poisons) a feedback mechanism causes the noncycling stem cells to reenter the cycle. Note that in radiotherapy the recovery of normal tissue may involve the reentry of temporarily noncycling cells into the cycle.

Permanently Noncycling Cells. These have left the cell cycle after having undergone a number of mitoses and reached a mature stage. They are represented by the *mature granulocytes* which, having lost their ability to reproduce, gradually die off (average normal survival about 10 days). They can be replaced only by cycling cells and temporarily noncycling cells.

Cycling cells must be undergoing mitosis, and therefore tumors and normal tissues containing large numbers of cycling cells must be more radiosensitive than those composed of noncycling cells. This accords with the Law of Bergonié and Tribondeau.

The *growth in volume* of normal and cancer tissue depends mainly on the net difference between the number of cells being produced and the number of cells being lost through death or out-migration per unit time. Terms used by radiobiologists in this context include cell cycle time, growth fraction, and rate of cell loss.

Cell cycle time is the time between successive mitoses, so the shorter the cycle time the faster the reproductive rate. Typical cell cycle times in normal human tissues range from about 40 to 220 hours and in human cancers from about 70 to 260 hours (Baserga, 1981).

Growth fraction pertains to the fraction of the cell population that is cycling, so the larger the growth fraction the greater the increase in the number of cells per unit time. The growth fraction of typical human cancers ranges from about 25 to 50 percent (Duncan and Nias, 1977).

The *rate of cell loss*, also known as the *loss factor*, is the fraction of cells that die or migrate out of the particular cell population per unit time. Typical values in experimental animal tumors range from 0 to 92 percent (quoted by Hall, 1978), and in human tumors 70 to 90 percent (Duncan and Nias, 1977).

Cell cycle times and growth fraction together determine the overall growth rate of normal and tumor tissue. Opposing this is the cell loss rate. If the net difference between these two trends is in the direction of an increasing number of cells, the tissue will grow in volume. In normal mature tissues and organs the rate of increase in cell number equals the rate of cell loss so the volume remains constant; the cell population is in *dynamic equilibrium*. In growing animals and in tumors, cell production exceeds cell loss, resulting in an increase in volume. Finally, cell populations composed of permanently noncycling cells, such as mature muscle and nerve tissue, ordinarily remain constant in volume; however, any mature cells that die cannot be replaced because there is no supply of potentially cycling cells.

Cell Population Kinetics

In some populations, such as the bone marrow, cells may move from primitive forms through intermediate stages of development, finally emerging as mature blood cells. In other populations no immature forms are present, so mature forms have to be replenished by reverting to less mature, cycling cells. In still other populations, mature cells can never be replaced once they have been lost.

The kinetics of various kinds of cell populations received a great deal of attention by Rubin and Casarett in their classical report (1968) on the pathologic changes induced in cells by ionizing radiation. They systematized the pertinent information as a basis for comparing the clinically observed radiosensitivity of various cells, tissues, and organs. Note that this expands the concepts of cell cycle and population kinetics, and follows the same definition of radiosensitivity, that is, the responsiveness of cells to ionizing radiation (analogous to the radiosensitivity of film emulsions).

We shall now summarize Rubin and Casarett's classification in which cells are divided into five categories, in decreasing order of radiosensitivity.

Vegetative Intermitotic Cells (VIM)—Class 1. Entering periodically into mitosis (e.g., cycling cells), these cells rank first in radiosensitivity (Law of Bergonié and Tribondeau) and have the shortest life span. They consist of primitive stem cells which, as mature cells of the same line die off, divide to produce daughter cells. Examples of VIM cells include hematopoietic (blood-forming) stem cells, epidermal (skin) basal cells, and ovarian and testicular primitive cells (oogonia and early spermatogonia, respectively).

Differentiating Intermitotic Cells (DIM).—Class 2. The VIM daughter cells go through several successive divisions with increasing differentiation between divisions. Hence, such cells are classed as *differentiating* (i.e., maturing) *intermitotic cells.* As their degree of differentiation increases, their radiosensitivity decreases. In addition, DIM cells have a slightly longer life span than VIM cells. Typical examples of DIM cells include the intermediate dividing and differentiating cells of a particular cell line such as the intermediate cells in the bone marrow (erythrocyte and granulocyte series) and the testis (intermediate forms of spermatogonia).

Multipotential Connective Tissue Cells (MCT)—Class 3. Three main types of cells occur in the connective tissue: endothelial cells, active fibroblasts, and mesenchymal cells. Their radiosensitivity lies about midway between the least and the most radiosensitive cell classes. Characterized by a variable life span, they can reproduce on demand as, for example, when they participate in the healing of injured tissue.

Reverting Postmitotic Cells (RPC)—Class 4. Such cells are relatively radioresistant. They do not ordinarily reproduce, but when injured or destroyed in sufficient numbers by various noxious agents, remaining mature cells can revert (go back) to a vegetative state and reproduce the same cells, as for example, in the liver. Other examples of reverting postmitotic cells include the parenchymal cells of the kidney, salivary glands, pancreas, endocrine glands, lung septal cells, and reticulum cells. In most cases the normal architecture of the organ is restored, but there are exceptions such as the liver. The reverting postmitotic cells have a relatively long life span.

Fixed Postmitotic Cells (FPM)—Class 5. Typically, these most-resistant cells have lost their ability to divide under any condition. They may be short or long lived, depending on the particular cell line. For example, the short-lived white blood cells (granulocytes) and cells at the tips of the intestinal villi (see Figure 8.04) have a short life span and cannot reproduce themselves but can be replaced only by activation of vegetative cells in their respective cell line. On the other hand, mature nerve and muscle cells have a long life span and once destroyed they can never be replaced because they have no available vegetative precursor cells. The fixed postmitotic cells belong to the class of permanently non-cycling cells.

Table 7.01 summarizes the characteristics of these five cell populations.

TABLE 7.01

RELATIVE CELLULAR RADIOSENSITIVITY.*

Cell Class	Radiosensitivity	Life Span	Degree of Maturity
vegetative intermitotic	+++++	+	+
differentiating intermitotic	++++	++	++
multipotential connective tissue	+++	+++	+++
reverting postmitotic	++	++++	++++
fixed postmitotic	+	+ to +++++	+++++

*Adapted from data of Rubin P, Casarett GW. *Clinical Radiation Pathology*, 1968.

As you can see, Rubin and Casarett have devised a dynamic concept of radiosensitivity in which the shifting relationships of dividing and developing cells serve as a basis for understanding the relative radiosensitivity of cell populations. By extension, we must realize that no cells are entirely radioresistant—tumors that appear to respond poorly to radiation actually are slow responders, requiring a relatively long time to undergo resolution, as well as large doses of radiation, in comparison with radiosensitive tumors. Table 7.02 indicates the relative radiosensitivity of various kinds of cells.

TABLE 7.02

DECREASING ORDER OF RADIOSENSITIVITY
OF MAMMALIAN CELLS (LETHAL ENDPOINT)*

Erythroblasts

Lymphocytes

Myeloblasts

Megakaryocytes (platelet precursors)

Spermatogonia (immature sperm cells)

Oögonia (immature ovarian cells)

Intestinal crypt cells

Basal cells of skin

Lens of eye

Gastric glands

Small blood vessels

Growing cartilage (chondroblasts)

Growing bone (osteoblasts)

Glandular epithelium

Renal tubular cells

Liver cells (parenchymal)

Connective tissue cells

Glia cells

Pulmonary alveolar cells   mainly secondary to fine

Mature cartilage   vascular and connective

Mature bone   tissue damage

Muscle tissue

Nerve tissue in
    brain and spinal cord

*Compiled from literature: Errera and Forssberg (1961, 1962); Rubin and Casarett in Dalrymple et al (1973).

In the next chapter, we shall take up the changes induced by radiation in various tissues and organs, predicated on the above classification. Later, the acute radiation syndromes will also be explained according to cell population kinetics and the relative radiosensitivity of the various kinds of cell populations in the human body.

Chapter VIII

RESPONSE OF NORMAL TISSUES AND ORGANS TO IRRADIATION

E xposure of humans to ionizing radiation is limited by the
tolerance of normal tissue. The term *tolerance* implies the
ability of normal tissues within the radiation field to recover
from any injurious effects. As you might suspect, this assumes
major importance in radiotherapy. However, radiation damage is
not limited to therapeutic doses; it may also occur at very low
dosage levels.

In this chapter we shall describe the early and late changes produced specifically in various tissues and organs by ionizing radiation
in doses higher than those in the diagnostic range, and shall also
present data on their tolerance levels. Later (Chapter XIV) we
shall deal with late somatic and genetic effects related to dose.

The major components of a typical organ include (1) the *parenchymal tissue* consisting of the functional cells peculiar to that
organ; (2) the *vasculature* or *blood supply*; and (3) the *connective
tissue*, which holds together the other components. Early radiation
effects involve mainly the parenchymal cells, although the endothelial lining cells of small blood vessels are also injured. Late
effects depend mainly on changes in vasculature and connective
tissue, resulting in fibrosis (scarring). Overall radiosensitivity —
susceptibility to radiation injury — depends on the interplay of the
radiation effects on the parenchyma, blood vessels, and connective
tissue.

Renewal Systems in Cell Populations

In Chapter VII we described the kinetics of the cell cycle and
cell populations. You should recall that the kinetics of cell popula-

tions deals with the reproduction of primitive or stem cells and their movement through successive stages of differentiation (maturation) to mature, functional cells.

Cell populations may be classified according to their particular *renewal system*, although there is no sharp dividing line among the various systems. By renewal system we mean the replenishment of mature cells from those in a more primitive stage. These renewal systems include (1) vertical or fast renewal, (2) horizontal or slow renewal, and (3) nonrenewal populations, which will now be explained and correlated with the terminology of Rubin and Casarett (see preceding chapter).

Vertical or Rapid Renewal Populations. Here, as mature cells normally age and die off, they are promptly replaced from the pool of primitive stem cells of the same cell line. This process is shown in Figure 8.01. The term *compartment* designates the various subpopulations in the cell renewal system. The stem cell compartment contains immature, functionless (i.e., vegetative intermitotic) cells which lie in wait, so to speak, and enter the *proliferating compartment* when called upon to do so by a feedback mechanism activated by loss of cells in the *mature compartment*. Some of the stem cells' daughters enter one or more successive dividing and differentiating (i.e., differentiating intermitotic cell) compartments where they become more and more differentiated, ending up as mature cells exactly like the ones they are replacing. Normally, the total number of cells remains virtually constant, the replacement rate being equal to the loss rate; this is called a *self-maintaining* system. Examples of vertical renewal systems include the lining cells of the intestinal tract (function in digestion and absorption of foodstuffs); various lines of bone marrow cells; skin (epidermis); and male gonads (testes). It must be emphasized that mature cells in vertical renewal systems cannot reproduce themselves—replenishment of such cells occurs by division and differentiation of precursor stem cells.

Horizontal or Slow Renewal Populations. Cell renewal occurs slowly and infrequently by reverting postmitotic cells, when there has been an unusual loss of mature cells. There is no available primitive or stem cell compartment, so that surviving mature

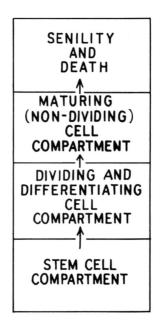

Figure 8.01. Vertical or rapid renewal population. Survival of the individual depends on prompt replacement of dying mature cells by proliferation of stem cells and eventual differentiation. Thus, a cell renewal system is at risk during the time interval between the death of mature cells and the arrival of replacement cells from the lower compartments. Examples include bone marrow cells, male gametes, and intestinal epithelium.

cells have to return to a vegetative (reproductive) state to replenish lost cells (see Figure 8.02). Included among horizontal renewal populations are vascular endothelium (lining cells of blood vessels), liver, and thyroid gland. For example, liver cells do not ordinarily reproduce, but if a part of the liver is removed, remaining liver cells begin to proliferate (reverting postmitotic cells). However, normal liver architecture is not restored, the new liver cells being arranged in the form of nodules. Another example is the thyroid gland whose mature cells can reproduce to form additional follicles under appropriate stimulation.

Nonrenewal Populations. Such cell lines (fixed postmitotic cells) have lost the ability to maintain themselves because they do not

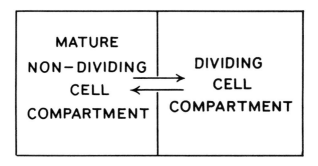

Figure 8.02. Horizontal or slow renewal population. No stem cells are available. When the loss of mature cells exceeds a critical point, remaining mature cells revert to dividing forms as represented by the dividing cell compartment from which the mature cell compartment is repopulated. Examples include the liver and the thyroid gland.

have a stem cell compartment, and mature cells cannot enter into mitosis. Examples include nerve cells, skeletal and heart muscle cells, and sense organs.

The promptness and severity of radiation injury depends on the *turnover time*, that is, the time interval between a stem cell and its derived mature cell. Obviously, injury to the primitive cells in a vertical renewal population such as the gastrointestinal tract will become apparent early because of the short life span of the mature cells and the time required for the stem cells to recover and enter the proliferating compartment. On the other hand, the liver, which typifies a horizontal renewal system, would take longer to evince radiation damage because the mature cells have a long life span, display less radiosensitivity, and have no available pool of radiosensitive precursor cells.

We shall next describe the response of various organs to ionizing radiation, classified according to the pertinent renewal system.

Vertical or Rapid Renewal Cell Populations

Bone Marrow. In adult humans, a low-level radiation dose of a large bone marrow volume causes suppression of hematopoietic (blood-forming) activity, manifested by depression of the peripheral blood counts. Active bone marrow in humans exists mainly in the sternum, ribs, and iliac crests, with smaller amounts in the ends of the long bones and in the vertebral bodies. The active bone marrow, usually referred to as *red marrow*, contains stem cells and immature forms of granulocytes (white blood cells), erythrocytes (red blood cells), and platelets in various stages of differentiation, as well as mature cells. These three blood cell lines are believed by most authorities to have a common stem cell, not shown in Figure 8.03.

Irradiation of the marrow depletes the stem cell population, eventually entailing a decrease in the mature elements (granulocytes, erythrocytes, and platelets) in the bloodstream. The maximum depression of cell count in each series depends on the turnover time of that particular series. Thus the granulocyte count drops first (in about two days), followed by the platelet count (about six days) and the erythrocyte count (about 110 days). These differences result from the different life spans of the mature cells in each series; only when they have decreased sufficiently in number, owing to the failure of the radiation damaged stem cells to replenish them, will their loss become apparent in the peripheral blood count. In other words, if we regard the stem cells as the "input" compartment and the mature cells of the three series as the "output" compartment, a decrease in the output compartment cells becomes manifest later if these cells have a longer life span. On this basis, we can explain the longer time required for the erythrocyte count to fall as compared with the granulocyte count, following bone marrow irradiation. However, it has been found that immature forms of erythrocytes are the most radiosensitive cells in the red marrow (see page 148).

As will be seen later, irradiation of the lymphopoietic organs which produce lymphocytes (lymph nodes, spleen, thymus gland, bone marrow) causes a much more rapid drop in the peripheral lymphocyte count because of prompt cell lysis (breakdown).

Gastrointestinal Tract. During a course of irradiation therapy wherein a segment of the gastrointestinal tract has been included in the treated volume, certain signs and symptoms usually ensue.

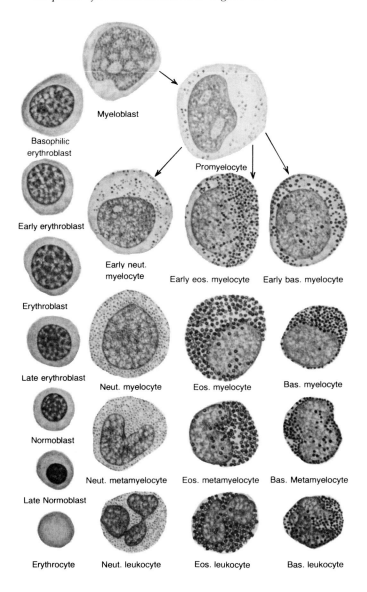

Figure 8.03. Normal development of human blood cells. The erythroblast and myeloblast are the earliest forms of their respective cell lines; they, in turn, most probably arise from a common stem cell. This is an excellent example of a vertical renewal population. (From W. Copenhaver et al. [Eds.], *Bailey's Textbook of Histology.* Copyright 1978, The Williams & Wilkins Co., Reproduced by permission.)

In this connection, the small intestine is the most radiosensitive part of the tract, although the stomach and colon do not entirely escape injury. Figure 8.04 shows the microscopic appearance of normal small bowel mucosa (lining). Primitive cells in the crypts at the bases of the villi cycle and gradually differentiate into mature cells as they move toward the surfaces of the villi.

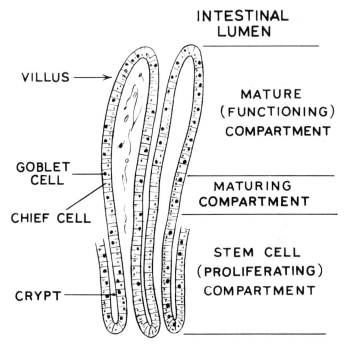

Figure 8.04. Microscopic appearance of villi (plural of villus) in the small intestine. The stem cells are located in the crypts; they proliferate and mature as they move toward the apex of the villus. As mature surface cells normally age and die off, they are replaced by proliferation and subsequent differentiation of crypt stem cells.

At approximately two weeks after the start of an irradiation therapy course in which the small bowel lies in the radiation field, "bowel upset" occurs, resembling nonspecific acute enteritis. Symptoms consist mainly of diarrhea, cramping, and tenesmus (straining). With larger doses, approaching 5000 rads (50 Gy) in fractionated doses, bleeding may occur. The sequence of events fundamentally involves injury to intestinal stem cells lying within the crypts (see Figure 8.05) from which normally develop the mature cells of the

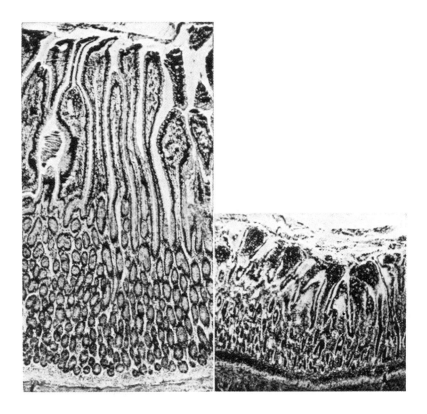

Figure 8.05. Effect of radiation on the small intestine of the dog.

A. Normal intestinal lining (mucosa), with the crypts at the bottom and the villi above.

B. Irradiated mucosa; marked damage to the crypt cells and flattening of villi, which have not been replaced by proliferation and maturation of crypt cells. (From A. Lacassagne and G. Gricouroff, *Action of Radiation on Tissues*, 1958. By permission of Grune & Stratton, New York.)

upper parts of the intestinal villi. Injury and death of crypt cells hinders replacement of the mature, functional lining cells, which are continually shed under normal conditions. As a result, the intestinal lining disappears, permitting the escape of fluid and

Elements of Radiobiology

blood into the lumen, and giving rise to diarrhea, cramping, and bleeding. If the radiation dose has not been excessive, surviving crypt cells soon multiply and differentiate into mature cells to restore an intact intestinal lining. With very large radiation doses, severe damage to blood vessels may cause subsequent ulceration, scarring, and stricture (severe narrowing) of the intestine.

Gonads (Reproductive Organs). In the male, the testis is made up of tubules containing spermatogenic cells in various stages of differentiation (see Figure 8.06). The younger intermediate forms

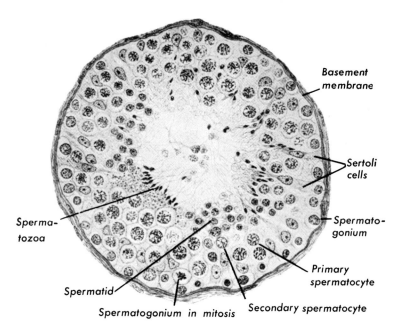

Figure 8.06. Normal development of human spermatozoa (sperm cells). This is another example of a vertical renewal population. The primitive cells are at the periphery, and as they proliferate (cycle) and differentiate, they move toward the center. (360× magnification) (From W. Copenhaver et al. [Eds.] *Bailey's Textbook of Histology.* Copyright 1978, The Williams & Wilkins Co., Reproduced by permission.)

(spermatogonia) display great radiosensitivity, so that a dose of about 500 rads (5 Gy) usually induces sterility. Figure 8.07 shows the depletion of cells in the testicular tubules following irradiation.

However, sexual potency does not suffer because the interstitial cells of Leydig which produce male sex hormones are relatively radioresistant and survive this dose. With fractionation, that is, a series of daily doses, sterility can be induced with a smaller total dose than with a single sterilizing dose. This is explained as follows: as the more radiosensitive spermatogonia die off, other nondividing spermatogonia enter the proliferating compartment, only to succumb to a subsequent dose fraction.

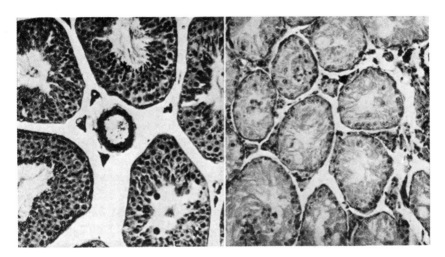

Figure 8.07. Effect of radiation on the testis of an adult mouse.

A. Normal testis (compare with Figure 8.06).

B. After irradiation in large enough dosage to cause sterilization, the spermatogenic cells have disappeared and so mature cells have also vanished. Only the resistant Sertoli cells remain in the tubules, and these interstitial cells have actually increased in number. (From A. Lacassagne and G. Gricouroff, *Action of Radiation on Tissues*, 1958. By permission of Grune & Stratton, New York.)

The *ovaries* (see Figure 8.08) respond differently to irradiation. Here the primitive germinal cells (the öogonia) have a fixed number at birth and decrease progressively during the woman's lifetime as an ovum matures each month. (In contrast, the spermatogonia continue to reproduce throughout the sexual life of the human male.) Irradiation of the ovaries destroys the primitive cells in the ovarian follicles thereby inducing sterility. Fur-

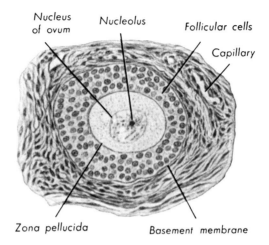

Nucleus of ovum　Nucleolus　Follicular cells　Capillary

Zona pellucida　Basement membrane

Figure 8.08. Primary follicle, an early stage in the development of a human ovum (egg cell). Each follicle gives rise to a single mature ovum (290× magnification). (From W. Copenhaver et al, [Eds.], *Bailey's Textbook of Histology.* Copyright 1978, The Williams & Wilkins Co., Reproduced by permission.)

thermore, the radiosensitive follicle cells manufacture female sex hormones so irradiation also induces premature menopause. Whereas a single dose of 500 rads (5 Gy) will induce menopause (permanent sterility) in one-third of women aged 30 to 35 years, it will do so in 80 percent of women approaching natural menopause. In the younger women, fractionation therapy raises the sterilizing dose to about 2000 to 3000 rads (20 to 30 Gy) for 100 percent effectiveness. Because of the uncertainty surrounding radiation sterilization and the possible survival of ova with radiation-induced mutations, it is ordinarily inadvisable to attempt radiation sterilization, although specific clinical indications may occasionally contradict this principal.

Skin. Careful consideration must be given to the skin during radiotherapy planning because the beam must pass through it to reach an underlying tumor. Skin represents a vertical cell renewal system similar to the gastrointestinal tract. The skin consists of the epidermis (epithelial cells) and dermis (underlying connective tissue). Cell renewal of the epidermis occurs by proliferation of the basal cells (see Figure 8.09). As the basal cells differentiate

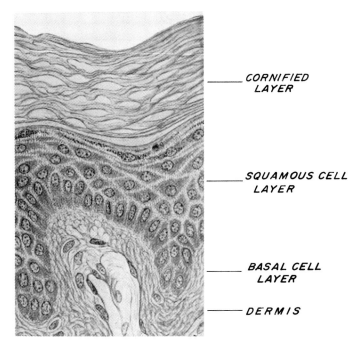

CORNIFIED
LAYER

SQUAMOUS CELL
LAYER

BASAL CELL
LAYER

DERMIS

Figure 8.09. Microscopic section of human skin. In this vertical renewal system, basal cells move toward the surface as they proliferate (cycle) and differentiate. Note that the intermediate cells are called prickle cells because of their apparent "spiny" surfaces resulting from prominent intercellular bridges. The surface squamous (flattened) cells produce a dry, cornified (hornlike) material. As the surface cells and cornified layer wear away, they are replaced by the underlying cells. (Adapted from W. Copenhaver et al, [Eds.], *Bailey's Textbook of Histology.* Copyright 1978, The Williams & Wilkins Co., Reproduced by permission.)

progressively toward the surface they gradually become flattened and keratinized (cornified), and are finally shed. Thus, replacement occurs continuously from the basal layer at a rate of about 2 percent per day (Duncan and Nias, 1977). Radiation injury, even with moderate doses, impairs mitosis in the basal cells thereby interfering with proliferation. In addition, capillaries and small arteries in the dermis are injured. With sufficiently large doses, especially in the orthovoltage range, the skin reaction passes through a sequence of erythema (reddening) and dry desquamation (shedding). With still larger doses, the reaction progresses to blistering and moist desquamation, and finally to ulceration (see Figure 8.10). Epilation (loss of hair) and diminished activity of oil and sweat glands may occur even with moderate doses. Healing takes place slowly, or not at all, with doses of sufficient magnitude.

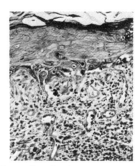 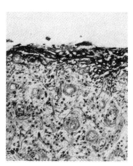

Figure 8.10. Effect of radiation on the skin.

A. Early radiodermatitis with reduction in thickness of the epidermis and destruction of the basal layer. An inflammatory reaction is present in the underlying dermis.

B. Reaction more severe four days later, with loss of epidermis (ulceration) and a more intense reaction in the dermis. (From A. Lacassagne and G. Gricouroff, *Action of Radiation on Tissues*, 1958. By permission of Grune & Stratton, New York.)

Megavoltage radiation, which typically delivers maximum dosage at some depth beneath the skin, causes the major reaction in the dermis or below; thus, excessive dosage in this energy region may cause subcutaneous (below the epidermis) woodlike brawny induration (hardening) of tissues associated with occlusion of blood vessels.

Severe late skin effects often consist of alteration in pigment, atrophy (thinning), and telangiectasia (spiderlike dilatation of superficial blood vessels). Ulceration is another late skin change, and this may eventually give rise to cancer. Late skin changes are uncommon with properly administered megavoltage therapy, although subcutaneous induration may occur when treatment is pushed to high levels in an effort to cure poorly responsive tumors.

The severity of a skin reaction depends on several factors. As would be expected, it increases with increasing radiation dose. However, daily fractionation, which permits recovery from sublethal damage during intervals between treatments, lessens the effect when compared to the same total dose given at one sitting. Furthermore, recovery between dose fractions improves as the interval between doses is lengthened, so that the total dose must be increased to achieve the same response. On the other hand, if one increases the size of the dose fractions, the response increases beyond that obtained with the usual dose fractionation. The so-called isoeffect curves for the time-dose relationship in skin therapy were first developed by Strandqvist as explained further on pages 246–250.

Mouth, Pharynx, and Esophagus. Moderate to severe effects are produced, depending on dosage and time factors. The sequence of radiation effects, especially in the pharynx and larynx, led Coutard (1932) to the principle of *fractionation therapy*, which is the bedrock of modern radiotherapy. For example, he observed (with orthovoltage therapy) an epithelial reaction or *epithelite* in the mucosa of the mouth and throat on the 13th day, and a reaction in the skin or *epidermite* on· the 26th day after the start of a fractionated radiotherapy course. The mucous membranes lining the mouth, pharynx, and larynx are less sensitive than those in the small intestine.

Typically, about two weeks after a course of irradiation has started, the oropharyngeal epithelium becomes reddened and swollen, soon to be followed by a white diphtheroid membrane. Pain and difficulty in swallowing, which may be quite severe, accompany the visible changes. Very large doses may give rise to ulceration which heals slowly or not at all. For example, 6000 rads (60 Gy) in six weeks gives about 5 percent chance of ulceration and subsequent severe fibrosis in the mouth and throat, and similar changes resulting in esophageal stricture.

Bladder. Radiosensitivity resembles that of the skin and oropharyngeal mucosa. During the usual course of therapy for carcinoma of the bladder or prostate, the bladder mucosa becomes grossly inflamed, and with large doses, approaching 6000 rads (60 Gy) in six weeks, may go on to ulcerate. Still larger doses, or shorter overall treatment time (in terms of days or weeks) may cause deeper reactions leading to scarring and contracture of the bladder wall. Such changes are dose and time related and are less pronounced with more extended fractionation (e.g. daily treatment to 900 rads per week instead of 1000 rads per week). The patient's symptoms include urinary frequency and burning. This clinical picture is often confused by the symptoms engendered by the tumor itself, as well as by attendant infection. Often, bladder pain results from blood clots rather than from irradiation effects; irrigation of the bladder with removal of the blood clots usually gives prompt relief of pain.

Horizontal or Slow Cell Renewal Populations

Blood Vessels. The lining endothelium is injured by moderate doses of radiation, certainly in the usual curative dosage range. Radiation-induced swelling of the endothelial cells causes blockage of capillaries and small arteries. This obviously reduces blood flow to normal tissues in the radiation field, thereby impairing the healing process. These vascular changes contribute greatly to radiation damage to the skin and other organs such as brain, spinal cord, kidney, and liver.

Growing Cartilage and Bone. Following an acute radiation exposure to immature cartilage and bone, no immediate changes occur. However, one eventually finds disordered development of bone in the damaged cartilage consisting mainly of impaired longitudinal growth of long bones and vertebral bodies (see Figure 8.11). For example, irradiating Wilms' tumor in infants and small children while including only one side of the vertebrae results in decreased vertebral height on the treated side with accompanying ipsilateral-leaning scoliosis. Similarly, irradiation of a wrist during its growth phase may cause inequality in the lengths of the radius and ulna by disturbing their relative growth rates, resulting in typical deformity at the wrist. A dose of 1000 rads (10 Gy) in one to two weeks has about a 1 to 5 percent chance of inducing growth arrest in growing cartilage and bone.

In the following organs radiation injury occurs mainly by effects on the *connective tissue*, especially the finer vessels and young fibroblasts, with relatively less direct injury to the parenchymal cells themselves.

Lungs. Pulmonary radiosensitivity is closely related to the total

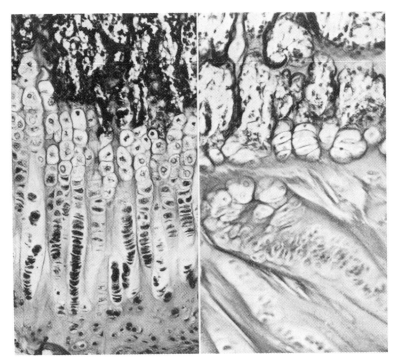

Figure 8.11. Effect of radiation on *enchondral ossification* (formation of bone on cartilage scaffold) in the rabbit. The normal, regular arrangement of cartilage cells is shown on the left. Irradiation effect on the right as evidenced by the disrupted pattern and abnormal cartilage cells. (From A. Lacassagne and G. Gricouroff, *Action of Radiation on Tissues*, 1958. By permission of Grune & Stratton, New York.)

volume of irradiated lung. With only a limited portion of a lung included in the radiation field, one finds vascular congestion (injury of endothelial cells) and accumulation of inflammatory cells (granulocytes, phagocytes, and red blood cells). These may be interstitial (in vascular tissue between alveoli) or alveolar (within pulmonary alveoli), or both. The parenchyma, that is, the alveolar cells proper, may also be injured. The term *radiation pneumonitis* encompasses the damaging effects of radiation on the lung. Symptoms range from minimal to severe. Typically, the patient complains of a dry, hacking cough; chest radiographs may reveal pneumonitis in the treated volume about one to two months after completion of therapy. Immediate response to corticosteroids is good in the face of mild injury, but severe pneumonitis may progress to pulmonary fibrosis. Irradiation of a single lobe or segment to a dose of 4000 rads (40 Gy) in four weeks has about a 5 percent chance of inducing pulmonary fibrosis. On the other hand, treatment of both lungs to a total dose of 2500 rads (25 Gy) in two and one-half weeks causes severe injury, characterized by pulmonary edema and right heart failure. It is, therefore, advisable in treating pulmonary lesions, as is equally true elsewhere, to shield as much normal tissue as possible. If a whole lung must be treated, the total dose should not exceed 2000 rads (20 Gy) in two and one-half weeks if one is willing to accept a 25 to 50 percent chance of inducing fibrosis.

Kidneys. Irradiation to moderate dosage causes a condition known as *radiation nephritis*. As in the lungs, organ damage here rests mainly on changes in the smaller blood vessels and connective tissue. Functional changes in the irradiated kidney can be detected by laboratory studies, which reveal a decrease in blood flow, and glomerular filtration and tubular excretion rates, even with doses as low as 400 rads (4 Gy) (Rafla and Rothman, 1974). With doses exceeding 2000 rads (20 Gy) in two weeks, after a latent period of six to 12 months, vascular injury results in hypertension, anemia, and cardiac enlargement. Severe inflammatory changes occur in the renal microstructure so that the kidney eventually becomes scarred and atrophic. Fatal outcome may be expected in about one-third of patients with severe bilateral radiation nephritis, although some may enter a chronic phase and die some years later as a result of hypertension and renal failure. The severity of the nephritis depends on the total dose and the fractionation schedule, and on the renal volume irradiated. Therefore, the kidneys should be so protected with a lead block during abdominal irradiation that the total dose to them does not exceed 2200 rads (22 Gy) in four weeks.

Major Salivary Glands. Of the three pairs of glands related to the mouth, the parotid gland is mainly responsible for secretion of the serous (watery) saliva. Radiation injury results from a direct effect mainly on the serous-secreting acini, with impaired production of watery saliva. The mucus-producing acini, being more resistant, continue to secrete a thick, ropy saliva. Complicated by radiation injury to the fine vessels and connective tissue, swelling of the gland and ducts contributes further to a drying of secretions. This condition, known as *xerostomia* or dry mouth, becomes apparent after an accumulated dose of about 3000 rads (30 Gy) in three weeks and is most uncomfortable for the patient. However it is unavoidable in treating certain lesions of the mouth and upper air passages. Surprisingly good recovery occurs in a number of patients despite intensive irradiation. *Severe* xerostomia has a 5 percent chance of occurring after a dose of 5000 rads (50 Gy) in five weeks, and a 50 percent chance after 7000 rads (70 Gy) in seven weeks. Not to be overlooked is the damage that may occur to the teeth, undoubtedly related to changes in the saliva. Typically, the necks of the teeth become blackened near the gum line and the teeth may actually loosen or break off.

Liver. Special care should be taken in irradiating the upper abdomen because the liver is much more radiosensitive than was once supposed. This holds true especially if a large portion of the liver lies in the radiation field. It is well to keep the total dose below 2500 rads (25 Gy) in five to six weeks owing to a significant chance of inducing radiation hepatitis at this level. Microscopic studies of livers so involved show severe congestion of the central zones of the lobules and hemorrhage from blocked small veins, although to a smaller extent there is also damage to the hepatic cells themselves.

Nonrenewal Cell Populations

Heart. Although the cardiac muscle has, in the past, been regarded as an extremely radioresistant organ, more careful observation of patients undergoing therapy for Hodgkin's disease and breast carcinoma has changed this view. It has been found that a fractionated course of radiation to a total dose of 4300 rads (43 Gy) in four and one-half weeks has a 5 percent chance of causing injury to the heart. Damage occurs almost exclusively to the connective tissue and small blood vessels, producing pericarditis and myocarditis. The myocardium (cardiac muscle) also suffers

direct damage with large doses but *cannot recover* because there are
no precursor cells from which the myocardium can be repopulated.

Central Nervous System. All available evidence incriminates
vascular injury by radiation as the predominant mechanism by
which nerve cells are damaged or destroyed. The sequence resembles that in other organs, with occlusion of smaller vessels by
swelling of endothelial lining cells, and formation of thrombi
(clots). The resulting loss of blood supply leads to necrosis (death)
of cells normally supplied by the damaged vessels. Owing to the
inability of mature nerve cells to divide (fixed postmitotic cells)
and the absence of precursor cells in the adult, the necrotic cells
cannot be replaced and a permanent nerve deficit results. The
clinical picture is complicated by the effects of the tumor itself. As
a rule, irradiation of the spinal cord beyond a dose of 4000 to 4500
rads (40 to 45 Gy) in 4½ to 5 weeks carries a high probability of
inducing serious injury characterized by transverse myelitis some
months after therapy. Hence, the spinal cord should be appropriately shielded to avoid exceeding this tolerance dose.

TABLE 8.01

RADIATION TOLERANCE DOSES OF VARIOUS ORGANS,
RANGING FROM MINIMAL ($TD_{5/5}$) TO MAXIMAL ($TD_{50/5}$).
BASED ON STANDARD TREATMENT SCHEDULE OF 1000 RADS/WK,
2 DAYS' REST/WK, PHOTON RADIATION ENERGY 1 TO 6 MeV.*

Organ	Injury at 5 years	1–5% $TD_{5/5}$†	25–50% $TD_{50/5}$‡	Volume or Length
Skin	Ulcer, severe fibrosis	5500	7000	100 cm^3
Oral mucosa	Ulcer, severe fibrosis	6000	7500	50 cm^3
Esophagus	Ulcer, stricture	6000	7500	75 cm^3
Stomach	Ulcer, perforation	4500	5000	100 cm^3
Intestine (small and large)	Ulcer, stricture	4500	6500	100 cm^3
Rectum	Ulcer, stricture	5500	8000	100 cm^3
Salivary glands	Xerostomia	5000	7000	50 cm^3
Liver	Liver failure, ascites	3500§	4500§	Whole
Kidney	Nephrosclerosis	2300	2800	Whole
Bladder	Ulcer, contracture	6000	8000	Whole
Testes	Permanent sterilization	500–1500	2000	Whole
Ovaries	Permanent sterilization	200–300	625–1200	Whole
Vagina	Ulcer, fistula	9000	>10,000	5 cm^3

TABLE 8.01 (continued)

Organ	Injury at 5 years	1–5% $TD_{5/5}$†	25–50% $TD_{50/5}$‡	Volume or Length
Breast (child)	No development	1000	1500	5 cm^3
Breast (adult)	Atrophy and necrosis	>5000	>10,000	Whole
Lung	Pneumonitis, fibrosis	4000	6000	Lobe
			2500	Whole
Heart	Pericarditis, pancarditis	4000	>10,000	Whole
Bone (child)	Arrested growth	2000	3000	10 cm^3
Bone (adult)	Necrosis, fracture	6000	15,000	10 cm^3
Cartilage (child)	Arrested growth	1000	3000	Whole
Cartilage (adult)	Necrosis	6000	10,000	Whole
Muscle (child)	No development	2000–3000	4000–5000	Whole
Brain	Necrosis	5000	>6000	Whole
Spinal cord	Necrosis, transection	5000	>6000	5 cm^3
Eye	Panophthalmitis, hemorrhage	5500	10,000	Whole
Lens	Cataract	250–500	1200	Whole
Ear (inner)	Deafness	>6000	—	Whole
Thyroid gland	Hypothyroidism	4500	15,000	Whole
Adrenal glands	Hypoadrenalism	>6000	—	Whole
Pituitary gland	Hypopituitarism	4500	25,000	Whole
Bone marrow	Hypoplasia	200	550	Whole
		2000	4000–5000	Localized
Lymph nodes	Atrophy	4500	>7000	—
Fetus	Death	200	400	—

*Data from Dalrymple, G. V., et al., *Medical Radiation Biology*, 1973. Courtesy of Saunders, Philadelphia.

†$TD_{5/5}$ is dose that, under these conditions, results in no more than 5% severe complication rate in 5 years.

‡$TD_{50/5}$ is dose that, under these conditions, results in a 50% severe complication rate in 5 years.

§These values may be too high especially with moving strip technique wherein doses of 2,450 to 2,920 rads in $2\frac{1}{2}$ weeks caused a high incidence of fatal radiation hepatitis. (J. T. Wharton. *et al.*, Am. J. Roentgenol. 117:73, 1973.)

Table 8.01 summarizes the radiation tolerance of various organs. Note the probability of injury as it relates to the total dose and overall treatment time.

Chapter IX

ASSESSMENT OF CELLULAR RADIOSENSITIVITY

Definition

Although various definitions have been proposed for the term *radiosensitivity*, it has generally come to mean the responsiveness of a tissue or tumor to ionizing radiation. From a medical standpoint, a tumor is said to be radiosensitive if it shrinks rapidly during a course of irradiation therapy. For example, a tumor disappearing early in the therapy course may be regarded as radiosensitive. Similarly, a normal tissue or organ that shows radiation injury after relatively small doses is said to be radiosensitive.

However, gross disappearance of a tumor does not necessarily signify that it has been completely destroyed, for viable cells may still remain as a nidus for tumor recurrence. In other words, a rapidly responding tumor is not necessarily highly radiocurable. In fact, a tumor that regresses slowly during radiotherapy may be more curable, in the long run, than one that regresses quickly.

A much more basic concept of radiosensitivity depends on the responsiveness of cells to ionizing radiation under experimental conditions, as manifested by *loss of reproductive ability*. We can see immediately the relevance of this concept in radiotherapy—once a tumor cell has lost its ability to undergo mitosis it is for all practical purposes dead, even though it may still be viable insofar as its nonreproductive functions are concerned.

In the following sections we shall cover the important aspects of cellular radiosensitivity: how it is measured, and what factors affect it.

Experimental Observations on Cellular Radiosensitivity

The demonstration by Puck and his co-workers (1956) that mammalian cells could be cultured artificially (similar to bacteria) opened the way to experimental radiobiology with subject material other than bacteria or plants. A mammalian cell growing in a suitable medium multiplies to produce a collection of cells called a *clone*. If a known number of cells are irradiated and then implanted in a culture medium, the failure of certain cells to reproduce and form a viable (i.e., capable of surviving) clone may be ascribed to radiation damage in these cells. Some radiation-damaged cells may die when they enter mitosis. Others reproduce only a few times but fail to generate viable colonies. In any case, *cell lethality in this sense means the loss of sustained reproductive capacity* of the cells in question, a phenomenon that has aptly been called *reproductive* or *mitotic death*. As noted above, reproductive death of tumor cells by ionizing radiation entails tumor eradication.

Because the loss of reproductive ability is a convenient measure of cellular radiosensitivity, it serves as an appropriate endpoint in experimental radiobiology. By general agreement, a cell is "dead" if it cannot form a clone of *at least 50 cells*, requiring more than five doublings.

Cell Survival Curves

Soon after Puck and associates (1956) showed how mammalian cells can be cultured artificially (*in vitro*), they discovered that as the radiation dose to a given number of cells is increased, a smaller and smaller fraction survives. This led them to the concept of a dose-dependent survival curve (analogous to survival of radioactive atoms with the passing of time). Cell survival curves represent the surviving fractions of cell samples exposed to various doses of radiation; the principle, shown in Figure 9.01, will now be explained.

Suppose we take a sample containing 50 cells (actually counted under a microscope) and implant them on a plate containing a suitable culture medium, and 40 clones (colonies) are produced in

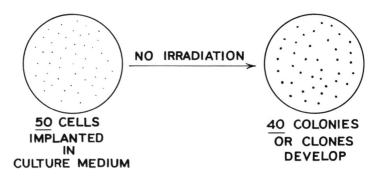

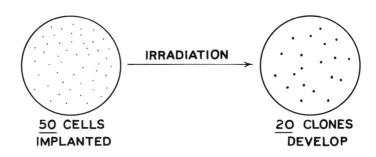

$$\frac{\text{NUMBER SURVIVING DOSE OF RADIATION}}{\text{NUMBER SURVIVING WITHOUT RADIATION}} = \frac{20}{40} \times 100 = 50\%$$

Figure 9.01. Principle of cell "survival" determination. Actually, survival in this sense refers to the ability of a cell to continue reproducing so as to form a *clone* of at least 50 cells.

the *absence* of radiation; this represents the normal cell viability in the hypothetical example:

$$\frac{40}{50} \times 100 = 80\%$$

Next we expose an identical sample of 50 cells to a small known dose of radiation and plate as before. If 20 clones result, the surviving fraction would be

$$\frac{\textit{no. surviving a dose of radiation}}{\textit{no. surving without irradiation}} = \frac{20}{40} = 50\%$$

The same procedure is then repeated for increasing doses of radiation. When the resulting data are plotted on a graph with the log surviving fraction on the vertical axis (dependent variable) and the dose in rads on the horizontal axis (independent variable) we obtain a survival curve.

The type of experiment just described, when carried out for a great variety of mammalian cells (except lymphocytes and oocytes) exposed to x or gamma rays, yields *survival curves* having the same general shape as shown in Figure 9.02. There is an initial *shoulder*

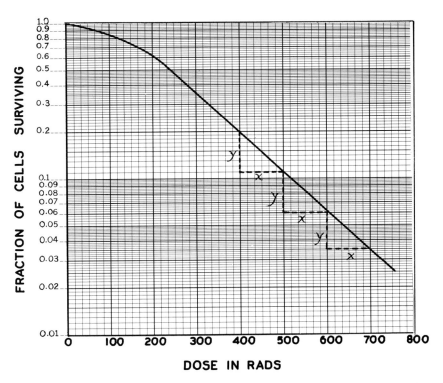

Figure 9.02. Typical mammalian cell survival curves. The initial *shoulder* portion represents *fast* or *Elkind recovery* of cells from low doses of radiation. The *straight* portion denotes that cell survival is reduced by the same *fraction* for *each equal increment* of dosage; damaged cells do not recover their reproductive ability in this dose range. Note that equal increases in dosage bring about the same *fractional* decline in cell survival. For example, for each 100-rad (1-Gy) increase in dosage (*x*) the fractional reduction in survival (*y*) is 0.55.

portion followed by a *straight* or *exponential* portion. Such a curve best fits a theoretical model derived mathematically on the assumption that cells contain *multiple sensitive "targets,"* a certain number of which must be "hit" and inactivated to cause cell death, that is, loss of reproductive ability (see pages 57–58).

It is of interest that *in vivo* methods of cell culture, such as bone marrow leukemic cells in mouse spleens and intestinal crypt cells and skin cells in mice, have made it possible to study dose-response effects. The resulting survival curves resembles those obtained with *in vitro* cultures (Hall, 1978).

Shoulder Portion. This represents the recovery of cells from *sublethal* (i.e., less than fatal) *injury* by low doses of radiation. Extrapolation (extension) of the straight portion back to the vertical axis gives the extrapolation number n as shown in Figure 9.02. The extrapolation number is one measure of the ability of cells to recover from sublethal radiation injury. In this multitarget single-hit model, n is thought to represent the number of sensitive targets in the cell. On the other hand, in a single-target single-hit model there would be no recovery and the survival curve would have no shoulder. Those cells that have not survived at low doses (shoulder portion) have been killed mainly by single hits, since they have little chance of accumulating sublethal injury in all n targets (see Withers in Fletcher, 1980, page 111).

Straight or Exponential Portion. This denotes the dose-response region in which a series of equal dose increments (increases) will entail a constant decrease in the *fraction* of surviving cells. For example, from the graph in Figure 9.02 we see that a dose increment of 100 rads, from 400 to 500 rads, reduces the surviving fraction from 0.2 to 0.11, or a reduction of about $0.11/0.2 = 0.55$; the next 100-rad dose increment 500 to 600 rads decreases the surviving fraction from 0.062 to 0.035, for a reduction of about $0.035/0.062 = 0.56$. In each case, equal increases in dosage give rise to a virtually equal *fractional* reduction in survival. In a semilog plot, equal intervals on the vertical axis represent equal fractions or multiples, whereas equal intervals on the horizontal axis represent equal differences or increments. The straight portion represents the summation of the effects of single and multitarget hits (Withers in Fletcher, 1980).

Implications of Exponential Response

Owing to this exponential response, more radiation is required to destroy a large tumor than a small one. Thus, if we were to start with 10^6 or 1 million cells and we knew that a particular dose would give a 75 percent survival along the straight part of the curve, 48 doses would be needed to reduce the surviving population to one cell (you can prove this by multiplying 1 million by 0.75, and then multiplying each successive answer by 0.75 until one cell remains—it would require 48 multiplications). But, if we start with 10^3 or 1000 cells, and the same dose as before yielded again a surviving fraction of 0.75, it would now take 24 doses to decrease the population to one cell. Thus, it takes twice as big a total dose (48 doses *vs* 24 doses) to reduce an initial population of one million cells down to one cell as it does an initial population only 1/1000 as large (i.e., 1000 cells) down to one cell. This relationship is shown in Figure 9.03.

An important fact about the straight portion of the curve is its increase in steepness with increasing cellular radiosensitivity. Looking at Figure 9.04, you can see that the same dose causes a greater reduction in the surviving fraction along the steeper curve (compare curve A with curve B).

Two important quantities have been derived from cell survival curves (see Figure 9.05). One is *mean lethal dose*, D_o, the dose that reduces the surviving fraction from a particular value on the straight portion of the curve, to 37 percent (0.37) of this value. Mathematically, D_o is the reciprocal of the slope (steepness of straight portion).

$$D_o = 1/slope$$
$$\text{or } slope = 1/D_o$$

We may conclude that the smaller the D_o value, the steeper will be the slope and the more radiosensitive the cell population under study. In other words, D_o represents an *inverse* measure of radiosensitivity at a cellular level. Of major interest is the fact that D_o has a narrow range of about 130 rads $\pm$ 50%, or about 65 to 195 rads for a wide variety of benign and malignant cells in *fully oxygenated* cultures.

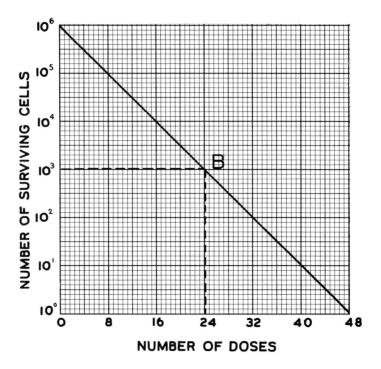

Figure 9.03. Relation between initial cell population number, and dose needed to reduce it to a particular level. For example, starting with a population of 10^3 (or 1000) cells, if 24 equal doses of x rays were needed to reduce the population to 10^0 (or one) cell (in straight line region of survival curve), then for an initial population of 10^6 (or 1 million) cells, twice as many or 48 doses would be needed to reduce the population to one cell. Thus, the number of doses is proportional to the *logarithm of the initial population*, for the same ultimate number of surviving cells. Note that the logarithm to base ten is the exponent of ten; in the above example, the exponents of ten are three and six, respectively, and the required number of doses is in the same ratio as 6/3.

The mean lethal dose D_0 should not be confused with the median lethal dose LD_{50} which is a measure of radiosensitivity pertaining to whole body irradiation of animals or other organisms; LD_{50} is the dose that causes death in 50 percent of irradiated population in a specified time period. Often this is expressed as $LD_{50/30}$, the dose causing death in 50 percent of an exposed population in 30 days (see pages 150–152).

Another important quantity derived from the straight portion

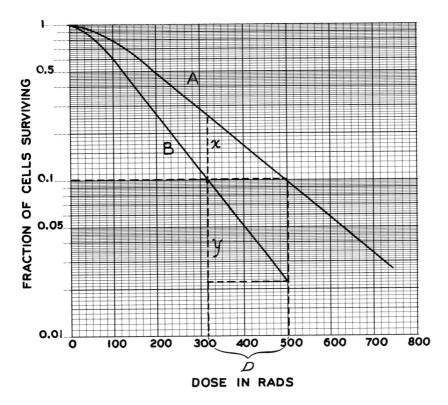

Figure 9.04. Curves showing that increasing radiosensitivity is indicated by a steeper straight portion of the cell survival curve. The same dose increment D causes a greater fractional loss in survival y in population B than the loss in fractional survival x in population A. Therefore B, with a steeper straight portion, is the more radiosensitive population.

of a cell survival curve is the *quasithreshold dose, D_q*, the dose at the point of intersection of the backward extrapolated straight portion with the 100 percent horizontal survival line (see Figure 9.05). Note that D_q represents the width of the shoulder portion in a cell survival curve, and indicates the degree of recovery from sublethal injury. From another standpoint, D_q indicates the radiation wasted before irreversible damage is sustained by cells. D_q also represents the difference between the sum of two equal radiation doses (separated by a few hours) and a single dose, which will engender the same decrease in the surviving fraction. For example, suppose

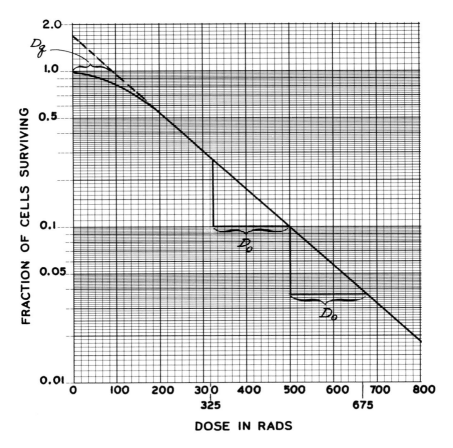

Figure 9.05. Cell survival curve showing the quantities *mean lethal dose* D_o and *quasithreshold dose* D_q. D_o is that dose which reduces the surviving fraction from a particular value (on the straight portion) to 37 percent of this value. D_o is the reciprocal of the slope (steepness), so the larger the D_o, the less the radiosensitivity.

D_q measures the width of the shoulder and therefore the ability of cells to recover from small doses.

the surviving fraction is reduced to 60 percent either by two doses of 175 rads (1.75 Gy) each or by a single dose of 300 rads (3 Gy). D_q would be obtained as follows:

$$2 \times 175 \ rads = 350 \ rads \ by \ split \ dose$$
$$1 \times 300 \ rads = 300 \ rads \ by \ single \ dose$$
$$D_q = 350 \ rads - 300 \ rads = 50 \ rads \ (0.5 \ Gy)$$

Here, too, D_q represents a measure of "wasted" radiation in the process of dose splitting (i.e., fractionation). For a particular cell population, the D_q value is the same whether obtained by measuring the width of the shoulder of the cell survival curve, or by finding the difference between the two-dose and single-dose values needed to reduce the population by the same fraction.

In summary, then, we may state that acute radiation injury to mammalian cells may cause reproductive death (loss of reproductive ability) and not necessarily actual death of the cells themselves. Cells undergoing reproductive death may proceed through one or more mitoses but cannot form a clone of at least 50 cells; many may die during the first few mitoses. Variation in the onset and severity of visible tissue or organ damage from radiation depends on the speed with which unirradiated or sublethally damaged cells can reproduce and replenish a cell population. This process is called the *kinetics of cell renewal*.

Effect of High-LET Radiation

We have discussed cell survival curves obtained in experiments with low-LET radiation (x and gamma rays), in which there occurs an initial shoulder that represents intracellular recovery from sublethal injury. With high-LET radiation such as neutrons or heavy charged particles, the shoulder decreases in width (i.e., smaller D_q). In fact, the higher the LET, the smaller will be the resulting shoulder, due to the increasing contribution of direct action of radiation on cells. High-LET radiation causes denser ionization tracks with increased probability of inactivation of multiple sensitive targets in the cell, and decreased probability of inducing sublethal injury. The curves in Figure 9.06 illustrate the typical differences between low-LET and high-LET radiation (i.e., x rays *vs* fast neutrons). Here you can see the very small shoulder and the steeper straight portion in the curve for neutron irradiation; the implication for fractionated dose therapy will be discussed later (see page 118).

Elements of Radiobiology

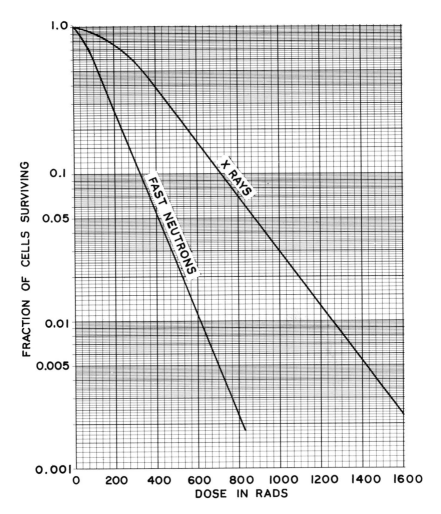

Figure 9.06. Cell survival curves comparing the effects of fast neutrons and x rays. Note the smaller shoulder and steeper straight portion with neutrons, indicating less recovery from small doses and greater radiosensitivity for this high-LET radiation.

Chapter X

FACTORS AFFECTING CELLULAR RESPONSE TO IRRADIATION

A number of conditions can modify the response of cells to irradiation, that is, alter their radiosensitivity. It must be emphasized at the outset that the operation of these factors depends to a great extent on whether the radiation is low- or high-LET.

While a great deal of theoretical and laboratory data have been accumulated, their impact on the practice of radiotherapy has been less than momentous. Here we shall focus our attention on those modifying conditions that have an important bearing on radiotherapy: (1) radiation type and quality, (2) fractionation and repair, (3) position of the cell in its reproductive cycle, (4) oxygenation, (5) physiologic state, and (6) chemical modifiers.

Radiation Type and Quality

With photon radiation—x and gamma rays—radiosensitivity *decreases* with increasing photon energy. For example, the absorbed dose of cobalt 60 gamma rays (av. energy 1.25 MeV) or 6-MV x rays must be increased about 15 percent above that of 250-kV x rays to yield equivalent biologic effects. In general, the principle of comparing the biologic response (i.e., biologic effectiveness) of various kinds of radiation is embodied in the *relative biologic effectiveness (RBE)*. (See also pages 31–34.)

RBE may be defined as the ratio of the absorbed dose of a standard type or quality of radiation to that of the radiation in question, to produce the same degree of a designated biologic effect. The standard has usually been 200 to 250-kV x rays with HVL 1.5 mm Cu. Thus,

$$RBE = \frac{absorbed\ dose\ of\ 250\text{-}kV\ x\ rays}{absorbed\ dose\ of\ test\ radiation} \qquad (1)$$

for the same radiobiologic effect.

Note that according to this equation, RBE depends on two factors: the type of radiation in the denominator, and the particular biologic effect under consideration, that is, the *biologic endpoint*. Accordingly, with a particular test radiation, RBE may vary with the stipulated biologic effect. Futhermore, RBE may not be the same for a particular test radiation under any and all conditions. It should be emphasized that the RBE of high-LET radiation *increases* with the degree of dosage fractionation (see page 118).

Because the average LET values of ^{60}Co gamma rays, megavoltage x rays, and orthovoltage x rays do not differ greatly (av. LET of ^{60}Co gamma rays is 0.3 keV/μ, and that of 200-kV x rays is 2 keV/μ), these three types of radiation are classified as low-LET. You should recall that photon radiation is indirectly ionizing—when it interacts with matter it first releases primary electrons which, in turn, cause a preponderance of ionization and excitation.

On the other hand, the average LET for 14-MeV *neutrons* is about 12 keV/μ, signifying that dense ionization tracks are produced. Recall that neutrons ionize indirectly by "knock-on" collisions with hydrogen atoms whose protons are set in motion and cause the ionization tracks. The ionization along these tracks is denser—more energy released per μ—and so we would anticipate a greater likelihood of multiple ionizing events within the cell, leading to inactivation of multiple sensitive targets (nuclear DNA), and greater biologic effectiveness than with low-LET radiation (see Figure 10.01).

With heavy, charged particles such as *deuterons* and *alpha particles*, LET is even higher—about 250 keV/μ for slow alpha particles. Such particles, ionizing directly and producing extremely dense ionization tracks, have an RBE of about 2 to 4.

In general, LET varies according to the *charge* and *speed* of an ionizing particle (see pages 31–32). However, the relationship is far from simple: the larger the charge the higher the LET because of the stronger associated electric field available for interaction with orbital electrons of atoms in the path of the radiation. On the

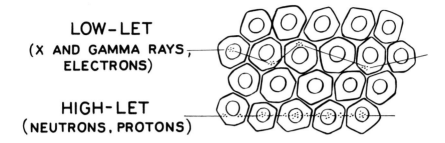

LOW-LET
(X AND GAMMA RAYS,
ELECTRONS)

HIGH-LET
(NEUTRONS, PROTONS)

Figure 10.01. Low-LET radiation releases ions and excited atoms (and molecules) cells apart. High-LET radiation releases them in multiple clusters within cells, thereby enhancing the likelihood of inactivating multiple targets (DNA) and so causing irreversible injury to the cell.

other hand, the slower the charged particle the greater the LET will be because the particle spends more time in the vicinity of the atoms in its path, thereby increasing the probability of interaction with orbital electrons. Thus, LET increases toward the *end* of the path of a charged particle as it slows down; therefore, LET is usually expressed as an average over the entire path of the particles under consideration.

It should not be assumed that RBE increases continuously with increasing LET. As LET rises to about 100 to 200 keV/μ for heavy ions, a peak RBE of about 8 is reached. Beyond this LET value, RBE falls because more ionizing events take place than are necessary to inactivate all the intracellular sensitive targets, so a part of the radiation is *wasted*. In other words, with radiation having an LET greater than the critical value of about 200 keV/μ its biologic efficiency actually decreases. The relationship between RBE and LET as well as the effect of oxygen, is shown in Figure 10.02.

Fractionation and Repair

According to the multitarget model, which best fits the response of mammalian cells to radiation, the cell survival curve derived from data on cell cultures has an initial shoulder, as already described (see pages 103–104). This shoulder represents the lower dosage range in which sublethally injured cells undergo recovery.

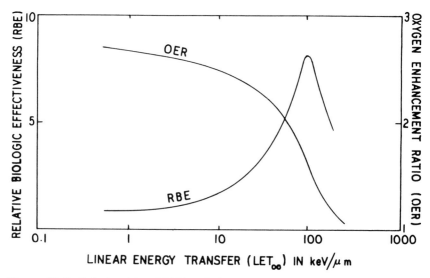

Figure 10.02. Relationship of RBE (relative biologic effectiveness) to the LET_{av} (linear energy transfer averaged over the entire track of the ionizing particle). Note that with low-LET radiation, which has a low RBE, the presence of oxygen greatly enhances radiosensitivity. The OER is much less (smaller oxygen effect) with high-LET high-RBE radiation. (*After Barendsen, 1968.*)

Further evidence for the ability of cells to recover from sublethal radiation injury derives from the observation that for a particular response in terms of surviving fraction, a larger total dose is needed if given in two equal fractions separated by at least two hours, than if given in a single dose. For example, we might find that 500 rads (5 Gy) in a single dose yields a surviving fraction of 40 percent, whereas 2 doses of 350 rads (3.5 Gy) each, separated by a time interval of two hours, would be needed to get the same surviving fraction. Thus, it would appear that by splitting the dose, a difference in dosage of 700 − 500 = 200 rads (2 Gy) had been "wasted," that is, an additional 200 rads (2 Gy) was required because cellular recovery had occurred during the time interval between the two doses.

Survival curves obtained with a two-dose model such as we have just described, in which at least a two-hour gap separates the doses, show reappearance of the shoulder in the second curve (see Figure 10.03). This means that when cells have recovered from a

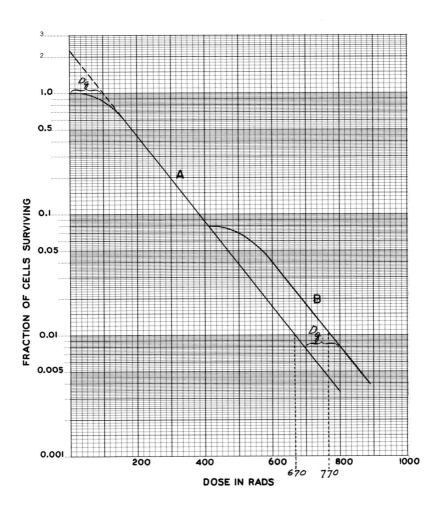

Figure 10.03. Characteristics of a two-dose cell survival curve. Curve *A* represents a typical survival curve obtained by exposing cell cultures to increasing doses of radiation. If the 10 percent surviving cells from a dose of 400 rads are now exposed to increasing doses of radiation, curve *B* is obtained having the same shape as curve *A*, with reappearance of the same shoulder (same D_q).

For a surviving fraction of one percent (0.01) a single dose of 670 rads would be needed as in curve *A*. With two doses separated by a gap of a few hours, 770 rads would be needed. D_q is the difference 770 − 670 = 100 rads. D_q is thus a measure of intracellular recovery from sublethal injury (Elkind or fast recovery).

relatively small dose of radiation, they respond to the next dose as if they had not previously been irradiated. If the split doses are identical the two curves have the same shape. Also, the straight portions are parallel (have the same slope) and are separated by a distance equal to D_q, the quasithreshold dose. This D_q is the same as the D_q for the first curve as shown in Figure 10.03. Thus, D_q represents the absorbed dose that must be accumulated by the cell population before the straight portion is reached; it is a measure of the width of the shoulder.

Recovery from sublethal radiation injury was discovered by Elkind and has appropriately come to be called *Elkind* or *fast recovery*. It results from intracellular repair (i.e., self-repair by the cell) during the time gap between two doses when the first dose was insufficient to inactivate all the targets within the cell. The exact mechanism for fast recovery is unknown. It has been suggested that fast recovery involves repair of single-strand damage to DNA, that is, to only one of the two strands in the molecule; however, this explanation has been contradicted by experiments that have failed to show correlation between single strand breaks and recovery from sublethal damage (see Withers in Fletcher, 1980, page 112).

Extension of the split-dose schedule from two doses to multiple doses becomes standard dose fractionation. Ever since Coutard's work (1932), irradiation therapy has consisted of a series of daily equal doses over total treatment periods measured in weeks or months. This kind of treatment schedule was evolved by close observation of many patients undergoing radiotherapy (see pages 93, 246).

Studies of dose fractionation with cell cultures have confirmed the observations on patients. Figure 10.04 shows, in a general way, the typical curves obtained with single dose, four-dose, and eight-dose schedules carried to the *same* total dose in each case. All doses are separated by an interval of several hours. Keep in mind that as the number of dose fractions increases, the individual dose increments must decrease to reach the same total dose. Now let us compare the three curves. As would be expected, the single-dose curve has a single shoulder and then becomes exponential (straight

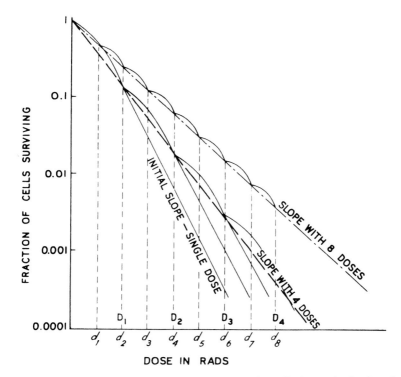

Figure 10.04. Modification of cell survival curves by splitting a single dose into multiple doses (extension of principle in Figure 10.03). Again, separation of doses by a time gap of a few hours results in reappearance of the shoulder (sometimes referred to as "return of *n*"). The larger the total number of fractions for the same total dose, the smaller the slope of the cell survival curve.

line on a semilog plot). With four doses there are four shoulders, and with eight doses, eight. Each shoulder represents Elkind or fast recovery from sublethal injury. Furthermore, the slope or steepness of the *resultant* straight portion becomes less as the number of fractions increases. Because the mean lethal dose D_0 varies inversely with the slope of the cell survival curve, we may conclude that as we increase the number of fractions (keeping the total dose constant) the slope decreases, D_0 increases, and radiosensitivity decreases. Note that this does not result from a change in inherent radiosensitivity, but rather from the larger number of recovery events associated with increasing fractionation.

More will be said later about fractionation in radiotherapy. At this juncture we should simply point out that the effects of dose fractionation have implications for the treatment of malignant disease. If fractionation were to decrease the responsiveness of both malignant and normal tissues to the same degree, no improvement in differential effect would accrue. However, if fractionation were to decrease the radioresponsiveness (i.e., enhance the recovery rate) of normal tissue relative to the tumor, an improved therapeutic ratio should result and, indeed, we see this in radiotherapy.

Thus far, we have limited our discussion to dosage fractionation and intracellular repair associated with low-LET radiation, that is, x and gamma rays. An important modification of response occurs with *medium-* or *high-LET* radiation. For one thing, higher LET radiation (e.g., fast neutrons) gives rise to survival curves having a small shoulder because the denser ionization more readily inactivates multiple intracellular targets, so that little if any repair of sublethal injury occurs. Furthermore, the straight portion of the curve has a steeper slope and a smaller D_0 than that for low-LET radiation (x and gamma rays) indicating greater radiosensitivity to high LET radiation (i.e. greater RBE). Comparative single-dose survival curves with x rays and neutrons are shown in Figure 10.05A.

High-LET radiation also changes the *shape* of the survival curve for fractionated dosage, again because of the virtual absence of fast recovery and the resulting very small shoulder. Comparison curves for fractionated doses of x rays and neutrons are shown in Figure 10.05B. Note especially how the RBE for neutrons increases with the degree of fractionation (compare RBE at 0.01 survival level in Figures 10.05A and 10.05B). Because of this, conventional fractionated dosage schedules with photon radiation must be carefully modified for high-LET radiation if serious radiation injury is to be avoided.

Position of Cell in Reproductive Cycle

Cellular radiosensitivity depends in large part on the position of the cell in the cell cycle (see Figure 10.06). As noted earlier, cells

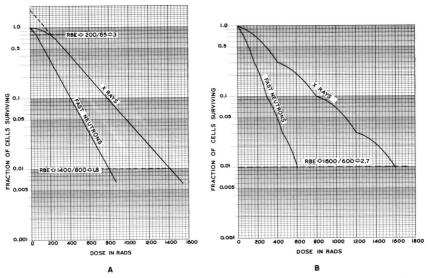

Figure 10.05. Comparison of cell survival curves for mammalian cells exposed to x rays and fast neutrons. In *A*, obtained with single doses, the shoulder is much narrower (smaller D_q) and the slope of the straight portion greater for neutrons than for x rays. Accordingly, the *RBE for fast neutrons becomes larger as the dose is decreased.*

In *B*, for fractionated doses in four equal fractions, the shoulder recurs after each dose, but again the shoulder for fast neutrons is narrower than for x rays; therefore, the *RBE for fast neutrons increases with increasing dosage fractionation. (Adapted from Hall EJ. Radiobiology for the Radiologist, 1978.)*

are most radiosensitive during mitosis, but there are peaks and valleys of radiosensitivity during interphase as shown in Figure 10.07. It turns out that most cells also show a peak of radiosensitivity during the G_2 phase.

In Figure 10.08 is shown a set of cell survival curves obtained by Sinclair (1968) with *synchronous* populations of Chinese hamster cells (all cells in same phase of cell cycle). These had been irradiated during various phases of the cell cycle. Cells in mitosis and in the G_2 phase display maximum radiosensitivity (steepest slope) and show no recovery (no shoulder). Absence of a shoulder in the survival curve is also consistent with single-hit inactivation of the "target." With this particular cell species the G_2 phase happens to be extremely short, so the curves for mitosis and G_2 virtually coincide. It should be emphasized that the G_2 phase of

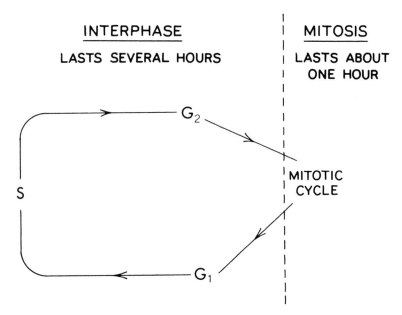

Figure 10.06. The Howard and Pelc (*Heredity*, 6:261, 1953) concept of a typical cellular generation cycle. There are two main phases—the synthetic S phase in which the DNA is normally doubled in amount, and the mitotic M phase in which the DNA is normally divided equally between the two daughter cells. The S and M phases are separated by two gaps, G_1 and G_2, in which other kinds of cellular activity occur, such as RNA and protein synthesis. S, G_1, and G_2 constitute interphase.

the cell cycle shows the greatest variation in time span from one species to another. This family of curves also shows that the late S phase is the least radiosensitive segment of the cell cycle.

Figure 10.09 presents the data of Terasima and Tolmach (1963) in graphic form. It shows the fraction of cells surviving a dose of 300 rads (x rays) to HeLa cells during various phases of the cell cycle. (HeLa cells are cultured human cervix cancer cells that have been re-cultured for many years.) Note again the maximum radiosensitivity (least survival) of cells in mitosis and G_2.

Obviously, if all the cells in a tumor could be synchronized so that they would all be in mitosis or G_2 at the same time, they would be in their most radiosensitive phases and there-

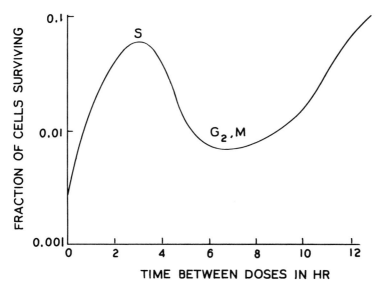

Figure 10.07. Effect of length of time gap between two doses (707 and 804 rads) on surviving fraction of exposed hamster cells. The surviving fraction is maximal when the gap is about two to three hours because the second dose has been given during the relatively less sensitive *S* phase. As the gap is increased to five to eight hours, the surviving fraction is at a minimum because the second dose has been delivered during the relatively sensitive G_2 and *M* phases. This indicates that it takes about five hours for the recovery shoulder to be lost. The cell population is now temporarily synchronized. *(Adapted from Elkind MM, Sutton-Gilbert H, Moses WB, Alescio T, Swain RW, 1965.)*

fore most vulnerable to lethal injury, with increased probability of tumor control. Unfortunately, experimental attempts to achieve synchrony have thus far been valueless in radiotherapy because the cells remain synchronized for only one, or at most two generations and then return to their former asynchronous state. Although radiation itself synchronizes cells, this too has a short duration.

Radiosensitivity differences manifested by cells in various phases of the cell cycle as just described for low-LET radiation also apply to higher LET radiation such as neutrons; that is, maximum radiosensitivity during mitosis and minimum radiosensitivity in

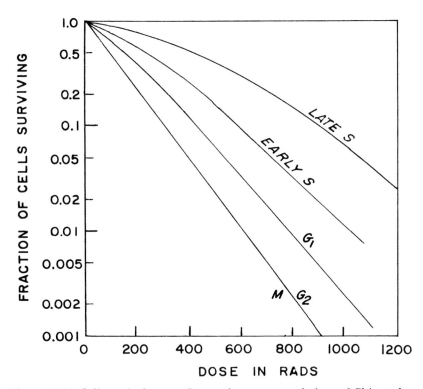

Figure 10.08. Cell survival curves for synchronous populations of Chinese hamster cells exposed to radiation during various phases of the cell cycle. Note the maximum radiosensitivity of the cells in mitosis (steepest curve) and absence of a shoulder (no recovery). The least radiosensitivity is shown by cells in the late *S* phase, accompanied by recovery at low doses (see pages 119–120). *(Adapted from Sinclair WK. Radiat Res 33:620, 1968.)*

the late *S* phase. However, the maximum *range* of radiosensitivity over the entire cell cycle is much narrower with neutrons than with x rays.

Oxygenation

Experimental evidence accumulated over many years has shown that the *radiosensitivity of cells increases with an increase in*

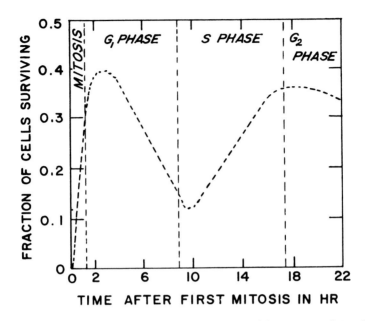

Figure 10.09. Fractional survival of HeLa (cancer of human cervix) cells in culture, exposed to x rays (300 rads) during various phases of the cell cycle. Maximum radiosensitivity (minimum survival fraction) is found in the mitotic (M) and late G_2 phases. Minimum radiosensitivity is indicated by the peaks in the curve (maximum survival fraction) during the G_1 and late S phases. *(Adapted from Terasima T, Tolmach LJ. Biophysical J 3:11, 1963.)*

oxygen concentration during the time of irradiation. This has come to be called the *oxygen effect.* However, it predominates only at low oxygen pressure up to about 30 mm Hg (mercury). As shown in Figure 10.10, the curve obtained by plotting radiosensitivity as a function of oxygen pressure rises very steeply as the oxygen tension (partial pressure) rises from zero to 30 mm Hg, then levels off abruptly. Thus, above this critical value, a further increase in oxygen tension has virtually no effect on radiosensitivity.

Figure 10.11 illustrates typical survival curves of cultured mammalian cells under hypoxic (low oxygen tension) and oxic (fully oxygenated) conditions. Note that the straight portion of the oxic curve is much steeper than the hypoxic. In fact, for any particular surviving fraction the dose required in this example is 2.5 to 3

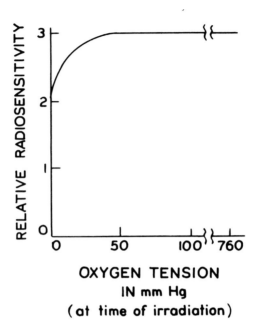

OXYGEN TENSION
IN mm Hg
(at time of irradiation)

Figure 10.10. Effect of oxygen tension (partial pressure) on radiosensitivity to low-LET radiation (x and gamma rays). As oxygen tension is increased there is an initial sharp rise in radiosensitivity. Above about 40 mm mercury (Hg) a further increase in oxygen tension has no additional effect on radiosensitivity. *(After Gray LH, 1953.)*

times greater under hypoxic conditions. Thus, oxygen behaves as a strong dose-modifying factor. Again note that to be effective it must be present at the time of irradiation.

The radiosensitizing property of oxygen is expressed numerically by the *oxygen enhancement ratio (OER)*, defined as the ratio of the dose in the absence of oxygen (anoxia) to the dose under full oxygenation, to achieve the same biologic endpoint:

$$OER = \frac{dose\ in\ anoxic\ state}{dose\ in\ oxic\ state} \text{ for same biologic effect} \qquad (2)$$

Typical OER values are 2.5 to 3 for low-LET radiation (x or gamma rays). With high-LET radiation, oxygen plays a much smaller role as a dose-modifying factor; so the OER for fast neutrons is about 1.5, and for alpha particles 1 (see Figure 10.12).

Thus, *the less the oxygen effect on radiosensitivity, the smaller the OER.*

In summary, OER indicates how much more radiosensitive cells are when fully oxygenated than when they are anoxic, for radiation having a particular average LET. The relation between OER and LET is assuming major importance in the future development of radiotherapy in the light of clinical trials with high-LET radiation such as neutrons, pions, and charged ions.

No entirely satisfactory or generally accepted explanation of the oxygen effect has thus far been proposed. One suggested mechanism focuses on the interplay between oxygen and free

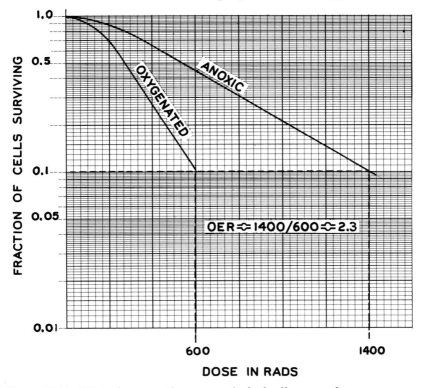

Figure 10.11. Effect of oxygenation on survival of cells exposed to x or gamma rays. Note the steeper (more radiosensitive) straight portion of the survival curve for oxygenated cells. Since OER = dose for anoxic cells/dose for oxic cells for the same effect, then to reduce the surviving fraction to 0.1, the dose for anoxic cells is 1750 rads, and for oxic cells 750 rads, giving an OER of 1750/750 = 2.5 for x and gamma rays.

Elements of Radiobiology

radicals during the radiolysis of water (indirect action of radia-tion). For example, oxygen may combine with the H· radical to form HO_2, and with the e^-_{aq} to form O_2, products known to be injurious to cells. Another pathway may involve the reaction of O_2 with an organic free radical R· in the target DNA molecule to form RO_2, an organic peroxide incapable of undergoing repair; thus, oxygen may "fix" the radiobiologic lesion (see page 61).

Physiologic State

In mammals various physiologic (functional) factors may in-fluence the radiosensitivity of the individual cells or the organism

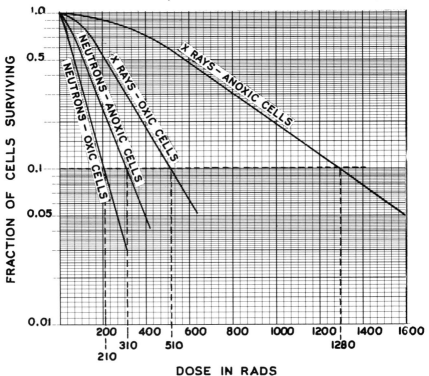

Figure 10.12. Comparison of OER for fast neutrons and x rays. For a 10 percent (0.1) survival, according to these curves, the OER for neutrons is 310/210 = 1.5, as compared with 2.5 for x rays. This confirms the greater sensitivity of anoxic cells to fast neutrons.

as a whole. These will now be mentioned briefly.

Age. The age of the animal at irradiation has an important bearing on response—embryos are more susceptible than mature animals of the same species because of the greater cellular radio-sensitivity in embryos (high mitotic rate and immaturity). For example, the LD_{50} (dose yielding 50% mortality) in mice increases nearly in proportion to the logarithm of the age up to about 13 weeks, then levels off, and finally decreases in old age.

Life-shortening. A number of experiments with small animals (Rotblat and Lindop, 1961) indicate that radiation causes a decrease in life expectancy, although there is some question as to the dose range required to produce this effect. At any rate, life shortening by irradiation becomes more significant the younger the animal at the time of exposure.

Endocrine Status. Certain endocrine (hormonal) abnormalities affect radiosensitivity. For example, hyperthyroid individuals display greater radiosensitivity than do normals, especially of their skin, although other tissues are not exempt from this effect. Also, female mice show less radiosensitivity than males.

Body Temperature. *Hypothermia* (subnormal temperature) decreases radiosensitivity by reducing the metabolic rate, whereas *hyperthermia* has the opposite effect. Studies are under way to determine whether such temperature effects can enhance tumor responsiveness relative to that of normal tissues in radiotherapy.

Nutritional State. In general, well-nourished individuals have a greater tolerance to radiation.

Chemical Modifiers

Radiation response can be modified chemically in two ways: enhancement and inhibition. These effects can be achieved, respectively, by *radiation sensitizers* and *radiation protectors*.

Radiation Sensitizers. A radiation sensitizer is any substance that increases the damaging effects of radiation. There are three kinds of radiation sensitizers: hypoxic cell and oxic cell sensitizers, and cytotoxic agents.

Hypoxic cell radiosensitizers include primarily *oxygen*. As we learned earlier, oxygen is probably the most potent chemical radiosensitizer, enhancing the lethal effect of low-LET radiation on cell cultures by a factor of 2.5 to 3. Certain chemicals called *hypoxic cell radiosensitizers* resemble oxygen by virtue of their ability to prevent repair of the damage inflicted by free radicals on the DNA macromolecule. However, they are more slowly metabolized than oxygen and therefore have more time to penetrate into the hypoxic regions of a tumor. At present nitroimidazole compounds such as metronidazole (Flagyl®) and misanidazole are undergoing serious clinical evaluation. However, their use is limited by toxicity to normal cells, especially nerve cells.

Oxic cell radiosensitizers have the ability to increase the responsiveness of already oxygenated cells. Included in this class of radiosensitizers are 5-bromodeoxyuridine (5-BUdR), 5-chlorodeoxyuridine (5-CUdR), and 5-iododeoxyuridine (5-IUdR), all of which are analogs of thymidine (precursor of thymine). Such analogs resemble the normal nucleotide (thymine or its precursor) sufficiently to "fool" the DNA-replicating process, which accepts the analog into the macromolecule. Their radiosensitizing action probably depends on a weakening of the sugar-phosphate backbone of DNA with a resulting increase in frequency of DNA strand breaks. Recall that single-strand breaks often repair themselves, but that double strand breaks tend to be permanent and therefore much more damaging to DNA. Unfortunately, the available oxic cell radiosensitizers have about the same effect on normal cells as on tumor cells. Of interest is the fact that even though 5-fluorouracil (5-FU), an analog of uracil, does not enter the DNA macromolecule, it is one of the more important chemotherapeutic agents against cancer.

Cytotoxic agents comprise various substances used in cancer chemotherapy. Among them are some which enhance the effects of radiation on cancer cells over and above their cytotoxic capability. Examples of such radiosensitizing cytotoxic agents are dactinomycin, methotrexate, and doxorubicin.

Radiation Protectors. Such compounds decrease the injurious effects of radiation on cells and include mainly the *aminothiols* or *sulfhydryls*. These contain an *SH* group (sulfur + hydrogen) near the end of the molecule so the *SH* group is relatively "free" and therefore highly reactive. An example is the amino acid cysteine from which a number of compounds have been derived and used experimentally. Protection turns out to be relative, not absolute. To be effective, such compounds, like oxygen, must be present at the time of irradiation. They lessen radiation damage to bacteria, plants, and various normal and malignant mammalian cells. Furthermore, they decrease the mortality from whole body irradiation in small mammals, particularly mice.

Protective action is maximal for low-LET radiation, and minimal for high-LET radiation (e.g., fast neutrons). A parallel situation exists for radiosensitizers — these give maximal enhancement of radiation injury with low-LET radiation, and least enhancement with high-LET radiation.

Several possible modes of aminothiol action have been proposed:

Competition with oxygen for free radicals released during radiolysis of water — the so-called *scavenger hypothesis*. Ordinarily, free radicals unite with oxygen to form *superoxides* which are extremely toxic to cells. The general form of a superoxide is RO_2^- resulting from the reaction of the free radical $R\cdot$ with O_2^-. Aminothiols interfere with this reaction by combining with free radicals before these can combine with oxygen.

Transfer of hydrogen to radiation-damaged DNA macromolecules, thereby promoting their repair.

Selective protection of normal tissues because these have more rapid blood flow than do tumors. Hypoxic cells experience less protection than normal cells in the presence of aminothiols.

Unfortunately, the usual aminothiol (SH) protectors are too toxic for medical use. However, certain *thiophosphate derivatives* of cysteamine have proved to be reasonably well tolerated by laboratory animals at effective dose levels, so these compounds are now undergoing clinical trial.

Chapter XI

THE IMMUNE MECHANISM AND RADIATION EFFECTS

When certain substances are introduced into the body, they activate a defense mechanism that either destroys or neutralizes the foreign substance. At times the response may be abnormal, excessive, deficient, or absent.

The term *immunity* refers to the body's ability to defend itself against various noxious intruders by calling on an *immune mechanism*. Reports in the literature have borrowed military terminology to designate various aspects of this phenomenon. For example, the entire immune process presupposes a *surveillance* system wherein watchdogs or sentries alert the system to activate *defenses* against the *foreign invaders*.

Immunity may be *innate* (inborn) or *acquired*. An animal has a *nonspecific* inborn defense mechanism represented by the skin and the lining of the respiratory and gastrointestinal tracts that act as natural barriers to the entrance of foreign substances. In addition, the body can acquire immunity that is highly *specific* for certain foreign substances called *antigens*.

What are antigens? They consist mainly of proteins, but certain complex sugars called polysaccharides also have antigenic properties, that is, the ability to incite an immune response.

Antigenic proteins may be living or nonliving. The former include viruses, bacteria, fungi, parasites, and foreign tissues and organs. Figure 11.01 summarizes antigens in diagrammatic form.

Antigens may gain access to the body by various routes: injection, ingestion, inhalation, skin contact, and tissue or organ grafting.

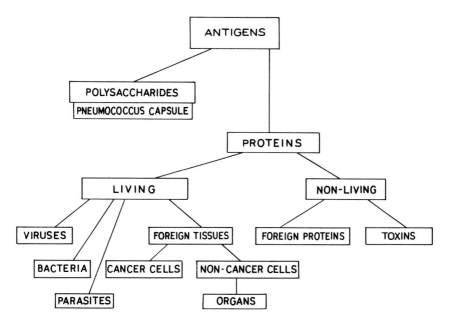

Figure 11.01. Classification of antigens.

The Immune System

Depending on the nature of the antigen, the immune response may be *humoral, cellular,* or *both.* Whatever the response, it requires the action of *small lymphocytes* of which there are two populations in the body: (1) *T-cells* which arise from stem cells in the bone marrow and mature in the thymus gland, and (2) *B-cells* which also arise from bone marrow stem cells but mature mainly within the bone marrow (possibly also in mammalian spleen and intestinal mucosa).

A third type of cell that participates in the immune response is the *macrophage,* a large cell that ordinarily resides in the connective tissue and has a strong tendency to engulf and digest foreign matter.

The body's immune system responds to the entrance of foreign protein or tissue, including organ transplants and antigenic tumors, by mobilizing B-cells, T-cells, or macrophages, alone or in various combinations. It is these elements which defend the body against the foreign antigen.

Immune Response to Tumor Cells

All cells have on their surfaces *specific antigens*, often called *markers*. These comprise the so-called *normal transplantation antigens* that account for the tendency of an animal to reject an incompatible tissue or organ transplant.

Surface antigens on cancer cells differ more or less from the normal antigen, and are therefore called *new antigens*. The individual harboring the cancer, through the immune system (probably by surveillance function of T-cells), recognizes that the new antigens differ from the normal transplantation antigens. However, both the new antigens and normal transplantation antigens are present on the cancer cell surface.

These new antigens are *tumor specific* in that they pertain to particular tumors, although other tumors may have identical or closely related antigens. The tumor-specific antigens originate mainly in cancers induced by viruses or chemicals, although they may occur even in spontaneous cancers.

How do tumor-specific antigens come into being? It is believed that carcinogenic viruses entering the cell bring to it "information" for the development of new antigen. Another mechanism may be the induction of changes in DNA, that is, a mutation by a chemical carcinogen.

Antigens of another type, the *tumor-associated antigens*, are not specific for particular tumors, but rather occur in a variety of tumors as well as in normal tissue. The *carcinoembryonic antigen (CEA)* exemplifies such tumor-associated antigens. The CEA is believed to represent carcinogenic information retained in the individual since the embryonic period but inhibited from being expressed until later in life. For example, there is a high incidence of CEA in human colon carcinoma; the amount of CEA in the body typically drops to a very low level after resection of the cancer and may rise to high levels when the cancer recurs. However, CEA is not entirely specific, as it may occur in other conditions such as infections. It should be realized that tumors marked by specific antigens may also have tumor-associated antigens on their surfaces.

As already noted, when transplantation or tumor antigens enter the body, its immune system calls upon the B-cells, T-cells, and macrophages to act as deterrents to bodily damage. Two main types of immune response may occur: (1) cellular, and (2) humoral.

Cellular Immunity. The T-cells, after maturing in the thymus, have been distributed in the lymphoid organs (e.g., lymph nodes, spleen). Foreign antigen such as transplantation or tumor antigen, upon entering the body, stimulates the T-cells to proliferate and become toxic to the tumor cells on contact with them. Such activated T-cells are capable of injuring or killing tumor cells in culture by releasing *cytotoxins* that produce holes in the cell membrane; they are therefore called "killer" T-cells.

The T-cells can be activated by still another mechanism, namely, by contact with *macrophages* that have been acted upon by tumor antigen. The macrophage itself may become cytotoxic as a part of the immune response, in cooperation with T-cells. Here the macrophage is attracted to a T-cell and activated by it. Such *activated macrophages* selectively kill cancer cells.

In attempting to transplant an organ such as a kidney, to avoid rejection by the recipient, we must use a histocompatible donor whose genetic makeup closely resembles that of the recipient. Identical twins are ideal for this purpose but are obviously not often available. In the event of histo*in*compatibility, the recipient must undergo *immunosuppression*, that is, must be rendered immuno*in*competent so that graft rejection will not occur. In the past, total body irradiation was used to inactivate the recipient's immune system, but this is now accomplished more satisfactorily by means of various chemicals.

Humoral Immunity. Here an antigen elicits the production of *neutralizing immunoglobulins*—specific proteins known as *antibodies.* How do these arise? B-cells, upon being activated by antigen alone, or by special "helper" T-cells, proliferate and develop into *plasma cells* that manufacture the specific antibodies. These antibodies react with particular antigens causing them to undergo agglutination (clumping) or lysis (dissolution). Insofar as tumor cells are concerned, antibodies specific for surface new antigens may damage or kill the cells on direct contact. Another mechanism involves the activation of another set of lymphocytes, the *K-cells*, which bind antibody and then kill the tumor cells.

We should know, at least in general terms, the stages of antibody formation because, as we shall see later, the effect of irradiation on the immune process depends on the instant of irradiation relative to the stage of antibody production. Figure 11.02 shows the basic phase sequence in this process. Note that after exposure to an antigen, there follows a latent period characterized by two phases. In the first or *preinduction phase*, T-cells join with antigen to stimulate B-cells, which then enlarge during the *induction phase*. When the B-cells become full-blown plasma cells, these produce the antibodies in the final or *antibody production phase*.

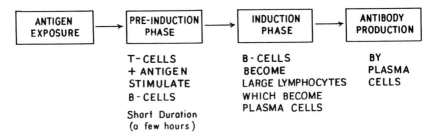

Figure 11.02. Stages in antibody formation.

The entire process is speeded up (i.e., shorter latent period) on subsequent exposure of the individual to the same antigen, the peak antibody level being higher than in the initial event. Thus, the individual's immune system "recalls" its previous experience and gets its guard up more quickly and in greater strength, a situation known as the *anamnestic* (remembering) *reaction*.

Effect of Irradiation on the Immune Response

Much of our knowledge about the effect of irradiation on the immune mechanism is derived not only from animal experiments but also from human irradiation preparatory to transplanting a genetically dissimilar organ.

In studies with small animals such as mice, rats, and rabbits, the radiation effect on immune response depends on when irradiation has occurred relative to the phase of antibody production.

In a classical paper by Taliaferro and others (1964) dealing with the hemolysin response of rabbits to injected foreign red blood cells (RBC) certain important observations resulted (see Figure 11.03). *Hemolysin,* an antibody formed in response to foreign (incompatible) *RBC* acting as antigen, causes lysis or breakdown of these RBC. Rabbits were exposed to a single whole body x-ray dose of 500 rads at various time intervals before and after injection of foreign RBC, and the rate and quantity of hemolysin formation measured. Irradiation *before* antigen injection causes a *decrease* in the rate and quantity of antibody production, the minimum being reached about one to two days after irradiation. Irradiation *after* antigen injection either has no effect on the quantity of antibody produced or actually increases it depending on the time interval; thus, irradiation given about one to two hours after antigen injection falls within the preinduction phase of anti-

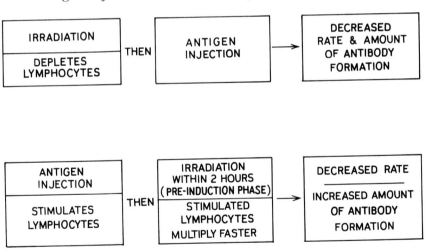

Figure 11.03. Effect of time of irradiation relative to antigen injection, on antibody formation. In the upper sequence, irradiation *before* antigen injection decreases the rate and amount of antibody formation.

In the lower sequence, irradiation *after* antigen injection (i.e., during the preinduction phase) causes an increased amount, but decreased rate, of antibody production. (*Based on data of Taliaferro WH, et al. Radiation and Immune Mechanisms, 1964.*)

body formation and actually increases total antibody production, although the speed of production decreases. Hence, even though the quantity of antibody may eventually rise above normal, the animal may still die because of the slowed rate of antibody formation.

We see, then, that radiation has an immunosuppressive effect if given *before* antigen. This has been explained by radiation-induced depletion of lymphocytes, which are ultimately responsible for antibody production.

On the other hand, when radiation is applied *after* antigen injection, some of the lymphocyte population will have had time to be stimulated by antigen during the time interval between the two events. Such antigen-stimulated lymphocytes, when irradiated during the preinduction phase, multiply more rapidly than the unstimulated ones and so have a greater ability to produce antibody and repopulate radiation-depleted lymphoid organs.

The preceding results were obtained with a single radiation dose. With fractionated irradiation schedules the immunosuppressive effect is less than that produced by the same total dose in a single exposure. This resembles the effect of fractionation in radiotherapy and in survival experiments with cell colonies, and suggests possible recovery of lymphocytes from sublethal injury.

Radiation also inhibits the phagocytic (cell-eating) activity of macrophages. This, together with the inhibition of antibody formation, makes the body susceptible to overwhelming infection by bacteria that ordinarily display little or no pathogenicity (disease-causing ability). As will be shown later (under Acute Radiation Syndromes) a sufficiently large dose of radiation destroys the intestinal cell lining allowing bacteria to penetrate the wall and enter the bloodstream. Escaping destruction owing to the deficient antibody formation and defective macrophage activity caused by irradiation, these bacteria are now free to bring about fatal septicemia (bacterial blood "poisoning").

Animals previously sensitized to a particular antigen retain their ability to respond later to the same antigen with normal antibody production even if the animal undergoes irradiation before antigen injection. However, the rate of antibody formation may be so slow that the animal dies before generating an adequate antibody level.

It has been known for many years that whole body irradiation causes a depletion of lymphocytes by its effect on the lymphoid tissue. Lead-shielding the appendix, lymph nodes, spleen, or Peyer's patches (in ileum) favors recovery because the shielded unirradiated lymphocytes repopulate the radiation-depleted organs. This, in turn, facilitates the recovery of the immune system, both humoral and cellular. In general, the ability to recover by this mechanism depends directly on the volume of lymphocytes protected by the lead shielding.

Human Cancer and Immunity

To varying degrees, the immune response to tumors may be cellular, humoral, or both. However, the antigenicity of human tumors is generally very weak, although Burkitt's lymphoma (African), melanoma, certain sarcomas, and infantile neuroblastoma, among others, do have some degree of antigenicity and can therefore induce cellular and humoral immune response.

The cellular immune response involves surveillance and recognition of the tumor by T-cells, which then proliferate and subsequently become activated by macrophages that have previously absorbed tumor antigen. As we have already mentioned, such activated T-cells are called "killer" T-cells (see page 133). The humoral mechanism also participates in the immune response to certain tumors.

In humans who are immunologically deficient for various reasons (heredity; immune suppression related to organ transplants; chemotherapy) we find an increased incidence of cancer. At the same time, we cannot explain why an antigenic cancer at its inception may escape surveillance and recognition by the T-cells and develop into a lethal cancer, despite normal immunocompetence of the individual. Relevant to this problem is the absence of an expected higher incidence of spontaneous tumors in immunoincompetent mice, that is, mice with impaired immune response. Such observations cast some doubt on T-cell surveillance as the most important factor in human immune response to tumors.

How can we, therefore, explain tumor suppression in the face of impaired immune response? The answer lies in the macrophage system. As we indicated earlier (see pages 133, 137), macrophages activated by previous absorption of tumor antigen, by certain substances such as endotoxins, or by BCG (a vaccine composed of weakened bovine tubercle bacilli) become killer cells for spontaneously arising tumors. Such killer macrophages have the peculiar ability to distinguish between normal and cancer cells, and release a necrosis factor that kills malignant cells. Thus, the activated macrophages possess an antitumor function independent of a specific immune response.

We have presented an extremely simplified version of tumor immunity, a complex subject involved in controversy and subject to modification at any time. Attempts are continually being made to use the immune response to tumors, weak though they may be, in practical therapy. This is still far from successful but may show promise in the future. Here are some of the therapeutic approaches under investigation:

Specific Immunity. This involves the induction of specific immunity in the individual against the tumor itself. Various products have been used in trying to induce specific immunity; these include tumor extracts, killed tumor cells, and live tumor cells that have been treated to enhance their immunogenicity.

Nonspecific Immunity. Numerous clinical trials are under way using agents such as *BCG* and *Corynebacterium parvum* to augment nonspecific antitumor immunity. This operates by stimulating general immunity, primarily by activating macrophages to become killer cells (see above). Specific tumor antigens and antibodies do not enter into this process. Neoplasms that have received such treatment, but with questionable benefit, include soft tissue sarcomas, malignant melanoma, and acute leukemia.

Transfer Immunity. It has been found that certain immune substances can be transferred from a donor individual to a recipient, conferring cellular immunity against the same tumor. Such immune substances include mainly *RNA* and *transfer factor* present in extracts of sensitized (immune) lymphocytes. Transfer factor, in essence, conveys specific immune capability from one tumor patient to another harboring the same type of tumor. When transfer factor is injected into normal animals it alters the macrophages so that they become active against the antigens to which the lymphocytes (from which transfer factor obtained) were sensitized beforehand in the donor. Transfer immunity still remains in the realm of experimental immunology and is not yet applicable in human tumor therapy.

Interferons. More recently, various types of *interferon* have been introduced and are being intensively studied for possible antitumor activity. However, the quantity and quality of the interferons are still inadequate for large clinical trials, although new methods of genetic engineering in which the DNA of bacteria is programmed to manufacture them, promise to improve the situation. Whether interferon of one type or another will turn out to be a potent anticancer agent remains problematical.

Chapter XII

ACUTE WHOLE BODY RADIATION SYNDROMES

U p to this point we have limited our discussion to the biologic effects of radiation on cell colonies and on limited volumes of tissue. We shall now take up the radiobiologic features of *whole body exposure* to a *single dose* of ionizing radiation as might be incurred in industrial accidents, or during the explosion of nuclear weapons. As will be brought out later, such exposure, usually with gamma rays, neutrons, or both, induces profound and often fatal effects on the victim. The promptness and severity of the injury depend on the absorbed dose. Damage from mechanical effects of nuclear weapons—the so-called blast—will not be included.

Whole body radiation shortens life to a degree depending on the dose up to about 1000 rads. Rotblat and Lindop (1961) found that in acute exposure of mice to various doses of radiation up to 800 rads, animals that recovered experienced premature aging and shortened life span amounting to about 5 percent per 100 rads. Examination of the body tissues in survivors revealed changes typical of those induced by irradiation of tissues and organs, namely, a decrease in the number of smaller blood vessels, loss of parenchymal cells, and increased density of connective tissue. However, radiation-induced aging is nonspecific. Although there is no direct proof, we would expect similar effects in humans, especially at the dose levels in experimental animals. In any event, whole body irradiation even at low dose levels may increase the incidence of leukemia (see pages 175–177).

Median Lethal Dose (LD$_{50}$)

We usually specify the lethal effect of whole body irradiation by the *median lethal dose (LD$_{50}$)*, the dose that causes 50 percent of the exposed population to die within a designated time period. Small mammals usually succumb within 30 days of exposure, although the peak mortality occurs at about 10 to 15 days. Therefore, the corresponding median lethal dose is expressed as $LD_{50/30}$. Human survival may last 60 days after exposure, with a peak mortality at about 30 days. So for humans the median lethal dose may be expressed as $LD_{50/60}$. We customarily regard a dose as being *immediately lethal* if death occurs before 30 days in small mammals such as rats and mice, and before 60 days in humans.

For the purpose of this discussion, we shall use exposure in R to express radiation quantity because of the difficulty inherent in estimating absorbed dose in many published studies. Furthermore, the stated exposures are at best approximate.

The lethal exposure to whole body radiation varies with the animal species. In man, the LD_{50} has been derived from the A-bombing experience in Hiroshima and Nagasaki (Japan) at the end of World War II. According to the U.S. Public Health Service, the following exposures typify the LD_{50} values in representative species:

Guinea pig	250 R
Dog	325
MAN	400
Mouse	530
Rabbit	800
Rat	850
(Bacteria	20,000 to 50,000)
(Viruses	50,000 to 1,000,000)

The LD_{50} is influenced by the state of health of the animals, the presence of bacteria or parasites, age, sex, and environment. Thus, the LD_{50} for guinea pigs approaches that for mice if disease-free guinea pigs are used in the lethality experiments. Young and old individuals display increased radiosensitivity (lower LD_{50}) compared to those in middle age. Also, males tend to be more radiosensitive than females. Finally, shielding a portion of the body such as the spleen or a limited volume of bone marrow reduces radiosensitivity (higher LD_{50}).

It is important to note that LD_{50} represents a small exposure, except for bacteria and viruses. In fact, the LD_{50} of 400 R in man would cause an insignificant rise in body temperature. Yet, much larger exposures may safely be given to localized regions in radiotherapy. As is well known, small exposures have been applied to the whole body (10 to 20 R) in treating lymphoma and chronic leukemia. It is estimated that, in humans, a total body exposure of 200 R gives a lethal probability of 1 to 3 percent, and an exposure of 600 R, 95 percent within 60 days. Obviously, lethal whole body exposures in humans would likely result only from accidents in nuclear facilities (e.g., reactors) or from atomic warfare.

Acute Radiation Syndromes

The totality of radiobiologic effects in the individual following an acute whole body exposure to a near-lethal or lethal dose of ionizing radiation is called an *acute radiation syndrome,* otherwise known as *radiation sickness.* This comprises a definite chain of events whose severity depends on the biologically effective dose as expressed in *rems.* For example, a particular absorbed dose of neutrons will produce a more severe biologic effect than an equal absorbed dose of photon radiation (x or gamma rays). Because of uncertainty in estimating the absorbed dose in total body irradiation, exposure in R will be used in this chapter. In general, the sequence of events is remarkably similar in various mammals.

Three main pathologic processes result from a large exposure to the whole body:

Necrosis or Cell Death. The most radiosensitive organs obviously suffer destruction first. Chief among these, in descending order of radiosensitivity, are (a) lymph nodes, (2) hematopoietic (blood-forming) organs, (c) gonads, (d) gastrointestinal epithelium, and (e) skin.

Hemorrhage. Bleeding occurs throughout the body, especially in the skin, gastrointestinal tract, respiratory tract, and other internal organs. Hemorrhages result principally from the *suppression of platelet formation,* the ensuing thrombocytopenia (low platelet count) interfering with the clotting of blood. Another factor is *damage to capillary walls* through which red blood cells can then more readily pass into the surrounding tissues. Finally, very high doses cause *inflammation* in and near small blood vessels.

Infection. This supervenes on account of the depletion of body defenses: loss of leukocytes, loss of lymphocytes with suppression of the immune mechanism, general debility, cell death, hemorrhage, and anemia.

We recognize four main forms of acute radiation syndromes, which depend on radiation dose and organ sensitivity. However, these syndromes overlap, one engendered by a larger dose having features of one induced by a smaller dose.

In the larger exposure range (above about 200 R) all the acute radiation syndromes have in common *four stages*, which differ in duration and intensity in the various syndromes:

1. *Prodomal Stage*—malaise, nausea, vomiting, fright.
2. *Latent Stage*—apparent recovery.
3. *Manifest Stage*—onset and progression of signs and symptoms of the particular syndrome.
4. *Recovery or Death*—depending on total exposure and treatment.

The *four acute radiation syndromes* will now be described individually and in increasing order of exposure. It must be kept in mind that the descriptions are based on animal experiments and on a few industrial accidents. In the latter, the radiation dose is often difficult to determine. Furthermore, published data on the various exposure ranges differ widely from author to author. We are including a reasonable estimate based on these published data.

Subclinical Syndrome

This may occur with a single whole body exposure of 50 to less than 200 R. Clinical findings include the *prodromal stage*: nausea vomiting, malaise, and fright. The leucocyte (white blood cells) count usually shows moderate reduction, with gradual recovery. Between 25 and 100 R one should not anticipate serious disability; there may be no symptoms, but transient slight depression of the leucocyte count may occur. Below 25 R there are no easily detectable changes, although special studies may reveal an increased incidence of chromosome breaks. The frequency of leukemia may also be increased in a population subjected to exposure even at this low level.

Hematopoietic Syndrome

When the whole body single exposure ranges from 200 to 600 R, death usually occurs in the second to third week, but may be delayed as long as two months. Note that the LD$_{50}$ (400 R) lies within this exposure range, so that mortality will be high. During the *prodromal stage*, which starts about two hours after the irradiation event and lasts about two days, the individual experiences nausea with or without vomiting. A *latent stage* follows during which the victim seems normal, but the bone marrow and lymph nodes are actually undergoing rapid cellular depletion. Following the latent stage, which may last two to three weeks, the individual enters the *manifest* stage of the syndrome, becoming seriously ill with a sore throat, fever, malaise, diarrhea, and petechiae (tiny hemorrhages in skin and mucous membranes). Circulating blood elements—lymphocytes, platelets, and granulocytes—rapidly decrease in number. Anemia occurs more slowly. Such depression of all blood elements is known as *pancytopenia*. At about three weeks after exposure, depilation (loss of hair) occurs. At lower exposure levels (near 200 R) hematopoietic recovery begins at about five weeks and is nearly complete in another three to six months. In the higher range (400 to 600 R) the latent stage is shorter, perhaps one week, more profound changes occur in the hematopoietic system, and a fatal outcome is much more likely in about two to six weeks because of hemorrhage and infection. With an exposure of about 400 R, about one-half the victims will die unless certain therapeutic measures can be instituted promptly, especially bone marrow transfusions from compatible donors, whole blood or blood element transfusions, and antibiotics. In general, with whole body exposures of 100 R eventual recovery may be expected. In the range of 200 to 600 R survival is possible, but exposures above 600 R lead to certain death in the absence of appropriate treatment.

Gastrointestinal Syndrome

With whole body exposures ranging from about 600 to 1000 R there is virtually 100 percent mortality in one to two weeks (average survival time six days). Onset of *prodromal symptoms*—nausea and vomiting—occurs in a few hours. In fact, prodromal onset in less than two hours makes it virtually certain that the exposure was in the neighborhood of 1000 R. Watery diarrhea may occur early. The *latent stage* is short, lasting a few days or, with large exposure, may be absent so that the prodromal stage passes directly into the *manifest stage*. This consists of nausea, vomiting, prostration, and steplike rise in body temperature. On about the sixth day, diarrhea becomes severe owing to the denudation of small bowel lining epithelium. Fluid now leaks into the bowel lumen through the bare surface and at the same time no fluid is reabsorbed, resulting in marked dehydration. Since bile salts cannot be absorbed in the distal ileum because of denudation, they pass into the colon, irritating it and intensifying the diarrhea. In addition, the denuded small bowel allows bacteria to enter the bloodstream through the bowel wall, causing sepsis and high fever, further intensifying the dehydration. During this time the hematopoietic organs manifest severe radiation injury: an initial rise in leucocyte count (cells released from storage depots) is soon followed by a fall in these elements as well as in the platelets. This results in uncontrolled *sepsis* (bloodstream infection) because there are no granulocytes to fight infection, and in intensified bleeding from the bowel because there are insufficient platelets to permit clotting of blood. Finally, the victim goes into shock and dies about one to two weeks after the initial whole body exposure. Note that although severe hematopoietic damage occurs in this syndrome, death results primarily from the severe injury to the bowel lining, which does not have enough time to recover. If exposure has been in the lower range, say about 700 R, some regeneration of small bowel lining cells can occur, particularly if the individual has been treated intensively with fluids, antibiotics, and other appropriate medication. However, hematopoietic failure occurs and results in death, unless the patient receives a bone marrow transfusion from a compatible donor, in addition to fluid replacement, blood transfusions, and antibiotic therapy.

Central Nervous System Syndrome

With a very high single exposure of several thousand R to the whole body, severe nausea and vomiting ensue within a *few minutes*.

The subject rapidly becomes dehydrated, drowsy, and lethargic, has difficulty with equilibrium (ataxia), and may experience generalized convulsions. The lethargy probably results from rapid onset of brain edema (swelling) and occurrence of inflammatory foci throughout the brain, whereas the ataxia and convulsions are associated with injury to the cells in the granular layer of the cerebellum. Severe vasculitis (inflammation of blood vessels) is also found throughout the brain. Death usually follows in a few days, but the precise cause remains uncertain; a major factor may be radiation damage to the nerve cells of the brain, either directly or as a consequence of vascular injury, or both. At the same time, the dose required to bring about the central nervous system syndrome causes severe injury to the gastrointestinal and hematopoietic systems, but death occurs before the associated syndromes have had time to become manifest.

Cell Population Kinetics as the Basis of Acute Radiation Syndromes

The hematopoietic and gastrointestinal acute radiation syndromes exemplify the principles of cell population kinetics in proliferating systems as described earlier (see pages 75–78). We are dealing here with the decline of a cell population after irradiation, followed by recovery or death.

Normally, in vertical or rapid renewal populations, as mature cells wear out and die off they are replaced by the proliferation of immature cells of the same cell line. Thus, in a normal state of equilibrium, when a sufficient number of mature cells die, a feedback mechanism signals the immature stem cells (i.e., vegetative intermitotic cells) to enter the reproductive cycle. Cells in the next and subsequent generations undergo progressive differentiation, ultimately becoming mature cells with their particular specialized function. Thus, the total number of cells tends to remain constant, the rate of cell replacement being equal to the rate of cell loss. We call this a *self-maintaining system*.

In Figure 12.01 we see in diagrammatic form the sequence just described for proliferating systems. The term *compartment* has been applied to the particular subpopulation of the cell renewal

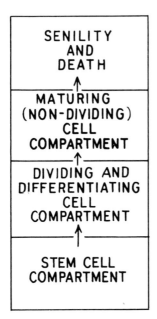

Figure 12.01. Vertical or rapid renewal population. Survival of the individual depends on prompt replacement of dying mature cells by proliferation of stem cells and eventual differentiation. Thus, a cell renewal system is at risk during the time interval between the death of mature cells and the arrival of replacement cells from the lower compartments. Examples include bone marrow cells, male gametes, and intestinal epithelium.

system. The *stem cell compartment* contains immature, unspecialized cells which start to reproduce when called upon to do so by the feedback mechanism activated by normal or abnormal loss of mature cells. Some daughter stem cells enter the *differentiating cell compartment* where they become differentiated or specialized for their particular function as mature cells, which then make up the *mature cell compartment*. Examples of such mature cells include the lining cells of the intestine and the various types of cells in the hematopoietic system. Pages 83–94 should be reviewed for details about these systems.

Let us now see what happens during irradiation of the hematopoietic and gastrointestinal systems with occurrence of the corresponding acute radiation syndromes. Both systems *normally* display a high rate of cell renewal, with continual loss of mature cells and rapid replacement by division and differentiation of immature cells. In other words, there is always a large fraction of stem cells and differentiating cells that are in mitosis, and hence in a *radiosensitive state*. Thus, in the acute radiation syndromes there is wholesale injury and destruction of the more primitive cells so that the dying mature cells cannot be replaced as rapidly as needed to maintain the integrity of the tissue or organ. When this occurs in life-sustaining tissue such as that found in the hematopoietic and gastrointestinal systems the individual dies. It is of interest that as early as 1968 Bloom found the hematopoietic stem cells actually to be less radiosensitive than the dedicated primitive cells—myeloblasts and erythroblasts.

Although the three types of blood cells have roughly similar differentiating and maturing times, they vary widely in life span as mature cells. Severe bone marrow depression appears within one week (hematopoietic syndrome), but granulocyte depletion becomes manifest in a few days. Therefore, repopulation of granulocytes, which are important in combating infection, must occur with sufficient speed for the animal to survive the hematopoietic syndrome. In the lower radiation dose range there may be enough viable stem cells remaining to undergo proliferation and maturation to provide adequate repopulation of granulocytes, but in the higher dose range this mechanism fails and the animal succumbs to the infectious process unless adequate treatment is given.

Platelets have a slightly longer survival than granulocytes, with a greater probability of repopulating before the individual dies.

Insofar as erythrocytes are concerned, their very long survival of 110 days means that anemia is late in appearing, and if the other cell lines can repopulate, the erythrocyte line has a chance to recover, especially if transfusions have been given.

Although the granulocytes are the first cellular elements in the peripheral blood to decrease in number following total body radiation, the immature cells of the erythrocyte line (red blood cells) are the most radiosensitive cells in the bone marrow. McFarland and Pearson (1963), and Adelstein and associates (1965), have observed that immature erythrocytes are the first to undergo an appreciable reduction in number after a total body dose of a few hundred rads. The number of reticulocytes (nearly mature erythrocytes that can be identified by a speckled appearance when stained with certain dyes) relative to the total red cell count in the peripheral blood is the most important indicator of depressed erythrocyte production after total body irradiation. Even though the total erythrocyte count is the last to fall after whole body irradiation because of their long life span, the erythrocyte precursor cells are the most radiosensitive. The next most sensitive are the granulocytes, and the least are the platelets in terms of the one-half reduction time, that is, the time it takes for the preirradiation count of that particular cell type to decrease 50 percent, as reported by McFarland and Pearson (1963).

In summary, then, relatively early death (two to three weeks) in the hematopoietic syndrome occurs from infection accompanied by a deficiency of granulocytes, and from hemorrhage caused by platelet deficiency; anemia appears later if the individual recovers, and as the erythrocyte population is replenished from viable stem cells, anemia gradually subsides.

It should be mentioned that lymphocytes arising in the common stem cell organs—bone marrow, lymph nodes, spleen, and thymus—are the most radiosensitive *mature* cells, disappearing before the other blood elements. Mature lymphocytes are readily killed by radiation despite the fact that they are not observed to undergo mitosis, this being an exception to the Law of Bergonié and Tribondeau. Lymphocytes in the above mentioned organs begin to die as early as one hour after irradiation, and those that survive show severe changes at six hours. Beginning replacement can be seen as early as five days, with well advanced recovery at two months. Recall that lymphocytes play an important role in the immune responses of the body, both humoral and cell-mediated. This undoubtedly contributes to the fatal outcome in whole body irradiation to which B-lymphocytes (bone marrow origin) are more sensitive than T-lymphocytes (maturing in thymus). Furthermore, the recovery of B-lymphocytes occurs in about one week, whereas that of T-lymphocytes takes about one month. Still, plasma cell antibody production does not stop immediately after irradiation but persists for several days.

In the *gastrointestinal syndrome*, involving as it does a vertical cell-renewal system, radiation injures primarily the *crypt cells* of the small intestine. However, the stomach and colon are also affected, but less severely than the small bowel. With midlethal exposures of about 400 R, necrosis of small bowel lining cells reaches a maximum at about 12 hours after irradiation and is accompanied by delayed mitosis in surviving cells. After several days, repair becomes evident. With exposures in the upper range— several thousand R—injury becomes so intense that long segments of bowel lose their cell lining. This leads to the clinical picture described above—diarrhea, dehydration, hemorrhage, infection, and sepsis—with virtually 100 percent mortality in one to two weeks. As would be anticipated, the hematopoietic syndrome also occurs at this exposure level, but death supervenes before this syndrome becomes full-blown, unless appropriate therapy is started promptly.

Finally, in the *central nervous system syndrome* we are dealing with nonproliferating (fixed postmitotic) tissue; adult nerve cells are highly specialized and do not reproduce. Besides, there are no stem cells to serve as a reservoir for the generation of new nerve cells. Thus, death does not occur, as in the other two acute radiation syndromes, from stem cell damage, but rather from indirect effects engendered by cerebral edema and radiation vasculitis and, probably to a lesser degree, from direct effects on the nerve cells themselves.

Chapter XIII

RADIATION EFFECTS ON EMBRYO AND FETUS

It has been known for many years that a dose of radiation delivered to an embryo during its early developmental stages causes much more severe damage than the same dose given to a mature individual. The radiosensitivity of the primitive cells in the embryo resembles, in a general way, that of similar cells in the adult. However, a relatively minor radiation insult to the embryo becomes amplified as development progresses, eventuating in severe congenital anomalies and growth retardation. Damage to embryonic stem cells, especially at lower dose levels, results from radiation changes in the cells themselves, rather than from vascular or connective tissue injury. When such injured primitive cells die or lose their ability to reproduce, abnormal development of the subsequent tissue or organ must certainly take place because not enough cells remain to develop a normal organ. In some instances, repair can still occur, resulting in a smaller though adequately functioning organ. Note that radiation to the embryo involves its whole body.

We must distinguish between radiation damage to the embryo's primitive cells with consequent organ abnormalities, and damage to ova and sperm that may later combine to produce an embryo. Radiation injury to the embryo itself gives rise to structural and functional changes in that particular individual, whereas genetic damage (i.e., to DNA in ova and sperm) is permanent and may express itself in future generations.

There is at present *no* evidence for a threshold dose, that is, a minimum dose below which no radiation injury occurs (see Figure 13.01). Radiation is injurious in any dose, but the severity of the effect is dose dependent because the number of cells damaged increases with the dose of a particular kind of ionizing radiation. The embryo as a whole is much more radiosensitive than the adult. Thus, Rugh (1963) states that the $LD_{50/30}$ for the embryo ranges from about 16 to 60 percent of that for the adult, which comes out to approximately 65 to 240 R (adult LD $_{50/60}$ is about 400 R).

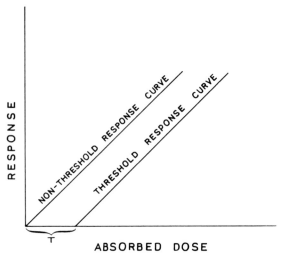

Figure 13.01. Dose-response curves for linear nonthreshold and linear threshold models. T = threshold dose, that is, the minimum dose that will evoke a response. No response occurs with doses below the threshold. With a nonthreshold dose-response model, any dose, no matter how small, will evoke a response.

By definition, in humans, the *conceptus* (product of conception) is called an *embryo* through the second month of pregnancy; thereafter it is called a *fetus*.

Radiation Effects During Gestation

By the term *gestation* we mean pregnancy, that is, the *act* of carrying or *period* of carrying the young from conception to delivery. Obviously, radiation experiments cannot be performed on pregnant humans. Radiobiologists must therefore rely on (a) experiments with small mammals, especially rats and mice, and (b) retrospective investigation of radiation accidents involving pregnant women as in the atomic bomb experience in Japan, as well as occasional instances when pregnant women received irradiation for a pelvic neoplasm. Unfortunately, we cannot depend on anecdotal reports in individual cases since at least 6 percent of live births in humans have anomalies unrelated to known sources of ionizing radiation, although at least some of these may have possibly resulted from background radiation.

What is the rationale for using small mammals for research on radiation injury to the embryo? It has been found that all mammals that have been studied respond similarly during embryonic development, organ for organ and stage for stage. Only the time table of the responses differs among various animal species, depending on their gestation period. For example, gestation in the mouse lasts 20 days, corresponding to 288 days in humans. However, the fraction of the gestation period occupied by various stages is not necessarily the same in mice and humans; thus, major organogenesis (organ formation) in mice lasts from about the sixth to the thirteenth day or up to *two-thirds* of gestation time as compared with the eleventh to the forty-first day or up to *one-seventh* of the gestation period in humans. In each different species we must compare radiation effects stage for stage.

Since the sequence of development in the human and mouse is the same within their respective time frames, the results of irradiating mice with various doses at selected embryonic stages can be extrapolated (carried over) to humans, to a reasonable degree of certainty. As a corollary to this, once we know from experimental evidence what effects are produced by a given radiation exposure during a particular stage of gestation, we can predict the outcome of such irradiation in humans.

In 1954 Russell and Russell, after intensive investigation of the effects of 200-R exposures in the mouse embryo, concluded that gestation may be conveniently divided into three stages: (1) preimplantation, (2) organogenesis, and (3) fetal development. We shall now describe these, extrapolated to the *human time frame* (see Figure 13.02); but keep in mind that damage to the embryo can occur with exposures as small as 10 R, or even less.

Preimplantation Stage in Humans (0 to 10 days). After fertilization of the ovum it usually takes nine days for the embryo to reach the uterus via a fallopian tube. This interval represents the period of preimplantation during which the embryo exhibits marked

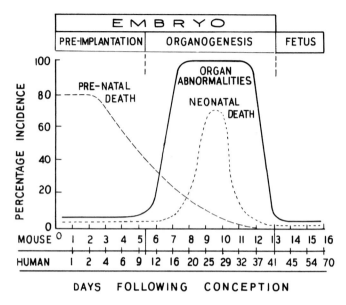

Figure 13.02. Effects induced by 200 R given on various days of gestation, with estimated comparative time scale for humans. Major organ damage occurs when radiation is given during period of major organogenesis; in humans this is about 11 to 41 days. (*Adapted from Russell LB, Russell WL. J Cell Physiol [Suppl 1] 43:103,1954.*)

radiosensitivity. In fact, *doses as small as 5 to 15 rads* may kill the embryo. However, surviving embryos are normal except for genetic chromosomal damage. In other words, irradiation during this very early stage has an all-or-none effect insofar as survival is concerned, but genetic effects may still occur.

   Organogenesis Stage in Humans (11 to 41 days). During the period of organogenesis, virtually every embryo surviving radiation exposure develops an anomaly. In general, the organ or tissue suffering the greatest injury is the one that was irradiated at the time that its precursor cells are undergoing the *most active multiplication (mitosis) and differentiation (maturation)*. Thus, in the mouse, brain herniation is especially prone to occur when the embryo is irradiated at eight days, nose and ear anomalies when irradiated at nine days, and microcephaly (small head) when irradiated at 11 days. In fact, the greatest *variety* of anomalies occurs in mice irradiated during the ninth to eleventh day of gestation, corresponding to the *25th to 27th day in humans*.

Murphy (1947) found in a study of women receiving pelvic radiotherapy during pregnancy, that out of 28 fetuses sustaining radiation injury, 16 (or about 60 percent) were microcephalic. Others had spina bifida, club feet, skull defects, hydrocephalus, or blindness, separately or in combination. Various authors using doses as small as 18 rads have found an anomaly rate of 25 percent. Table 13.01, adapted from Rugh (1965), gives the day of gestation in man (extrapolated from the mouse) when irradiation will induce a peak incidence of a particular anomaly.

TABLE 13.01

EMBRYONIC OR FETAL AGE AT WHICH
A PARTICULAR ORGAN IS MOST SUSCEPTIBLE
TO RADIATION DAMAGE IN HUMANS
(EXTRAPOLATED FROM MOUSE DATA).*

Anomaly	*Human Gestation Days*
embryonic death	4–11
exencephaly	0–37
cataracts	0–6
anencephaly and microcephaly	9–90
cleft palate	20–37
cardiac	21–29
spina bifida	16–54
anophthalmia	16–32
microphthalmia	20–54
skeletal	25–85
growth disorders	54+
equilibrium disorders	37+

*Data selected from Rugh, R. *Radiology* 99:433, 1971.
  Note that the 20th to the 32nd day of gestation includes the period of maximum susceptibility to congenital anomalies of all kinds.
  In general, 25 rads to the embryo during major organogenesis is definitely teratogenic.

In humans, central nervous system malformations comprise an especially important outcome of irradiation *in utero*. In their most primitive stage (neurectoderm) nerve cell precursors are moderately radiosensitive, requiring about 400 rads to kill all cells (Rugh, 1965), in contrast to a dose of about 10,000 rads for the same effect in mature neurons (nerve cells). However, from the 17th day of gestation to full term delivery (and a few weeks into the neonatal period) neuroblasts are present throughout the central nervous system; a dose of only 25 rads is lethal for these extremely radiosensitive cells, showing why nervous system malformations represent the most common type of radiation anomalies in embryonic and fetal life.

Fetal Development Stage in Humans (6 weeks to term). During this stage, visible malformations are mild or absent, depending on dose. The lethal dose approaches that in adults. Functional impairment does occur but is difficult to identify, involving mainly disorders of intelligence and growth.

Dekaban (1968) has summarized the anomalies most likely to occur in human embryos and fetuses exposed to 250 rads at various times during gestation:

Less Than 2-3 Weeks—high probability of lethal effect with absorption of embryo; low probability of anomalies.

4-11 Weeks—severe anomalies of many organs in most children.

11-16 Weeks—microcephaly, stunted growth, and genital organ anomalies.

16-20 Weeks—mild microcephaly, mental retardation, and stunting of growth.

More Than 30 Weeks—rare visible abnormalities, but functional impairment such as mental deficiencies may occur; these may be too difficult to demonstrate. Midlethal dose approaches that in adults.

Application in Medical Practice

Because of the lack of evidence for a threshold or minimum harmful dose, as well as the pronounced susceptibility of the human embryo to ionizing radiation, we should keep diagnostic x-ray and nuclear medicine procedures to the minimum consistent with medical needs, especially in women of childbearing age. It is of particular importance that during the most radiosensitive period—first six weeks of pregnancy—the mother is usually unaware that she is pregnant.

Accordingly, some authorities have recommended that any radiodiagnostic procedures that may expose the uterus of women in the childbearing age should be postponed, if medically feasible, to the so-called "safe" period, that is, within the first ten days after the beginning of a menstrual period. By scheduling such examinations during the "safe" period we can presumably avoid risk to an unsuspected embryo.

Why is the "safe" period so regarded? The human ovum (egg) is released from the ovary approximately 14 days after the onset of menstruation (see Figure 13.03). If fertilization of the released ovum does not occur, it is obvious that no pregnancy can result. The ten-day "safe" period has been selected to allow for variation in the time at which the ovum is expelled from the ovary, that is, to provide a margin of safety. Accordingly, to avoid radiation injury to an unsuspected early pregnancy, examinations of abdominal organs, hip, lumbosacral spine, and pelvis should be limited to the "safe" period insofar as possible.

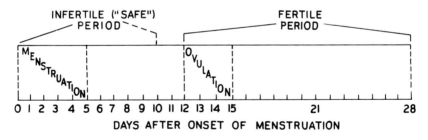

Figure 13.03. Schematic representation of human menstrual cycle. The sequence is not so precise in all women, and during different cycles in the same woman, giving rise to error in estimating "safe" period—the time interval between onset of menstruation and ovulation.

On the other hand, most authorities in this country now believe that the ten-day rule gives a false sense of security because there is too much variation in the time at which the ovum is released from the ovary. Furthermore, it makes more sense to limit diagnostic x-ray and radionuclide procedures in potentially pregnant women to those that are medically necessary. In this case, we must, if possible, collimate the beam and shield the pelvis with lead, and carefully restrict the number of film exposures.

X-ray Pelvimetry

In recent years the use of pelvimetry has declined as obstetricians have become aware of the potential radiation hazard to the fetus. Yet, the estimated pelvimetry rate remains at about 6 percent of live births (Kelly, 1975; Campbell, 1976).

What is pelvimetry? It is a radiodiagnostic procedure that measures the three important levels in the maternal pelvis: inlet, midplane, and outlet. It is ideally supposed to determine whether the child can be delivered without a cesarian section. The procedure usually requires the exposure of two films, which deliver an average fetal dose of about 620 mrads with a medium speed imaging system (UNSCEAR, 1977).

The question has arisen as to the validity and place of pelvimetry in the management of labor, and so a number of experts have turned their attention to this problem. Thus, Kelly et al. (1975) found that pelvimetry, the largest single source of fetal irradiation, rarely changes the obstetrician's decision as to whether a cesarian section is necessary. Furthermore, there is little agreement as to the indication for ordering this procedure. Joyce et al. (1975) decided in their study that pelvimetry during pregnancy is rarely justified except in breech presentations, and even then it should not be done routinely. Campbell (1976) found only one or two really small pelves among 2500 patients.

Based on the deliberations of a panel of experts, the U.S. Department of Health and Human Services (FDA, 1979) issued the following recommendation:

Pelvimetry is not usually necessary or helpful in making the decision to perform a cesarian section. Therefore, pelvimetry should be performed only when the physician caring for the patient feels that pelvimetry will contribute to the decisions concerning diagnosis or treatment. In those few instances, the reason for requesting the pelvimetry should be written on the patient's chart. The statement does not apply to x-ray examinations for purposes other than measurement of the pelvis.

This was approved and adopted by the American College of Radiology. A comparable policy statement was issued by the American College of Obstetricians and Gynecologists.

When x-ray pelvimetry is ordered, careful technic must be used to avoid repeat examination. Although as stated above, a two-view pelvimetry delivers about 620 mrads to the fetus with medium speed films and screens, the fetal dose varies with the radiographic equipment and screen-film combination.

This dose can easily be reduced to 150 mrads by the use of rare earth screens. As we shall see later (page 176) there is good evidence that abdominal irradiation of pregnant women increases the incidence of cancer, especially leukemia, *by* 40 to 50 percent within the first 15 years of the child's life (McMahon, 1962; Kneale and Stewart, 1976).

Chapter XIV

LATE EFFECTS OF IONIZING RADIATION

W e may conveniently discuss the delayed harmful effects of ionizing radiation under two headings, according to the dose received: (a) high doses in the range of a few hundred rads, and (b) low doses amounting to a few rads. Common to both is the existence of a *latent period* between the actual exposure and the appearance of the effect. This delay may extend from a few hours to one year for *early effects*, and from one year to many years for *late effects*. Furthermore, radiation injury may occur after a single, or after repeated or chronic, exposure. (The acute whole body radiation syndromes were described on pages 142–146.)

As we have already learned, radiation injury may result from either localized or whole body exposure, the intensity of the effect depending on the radiation quality, total dose, and overall time during which radiation exposure occurred. The likelihood and nature of localized effects will depend also on the region of the body involved, since organs vary in radiosensitivity (see pages 80–99).

There are two main categories of effects depending on how they manifest themselves. These are *stochastic* and *nonstochastic*.

Stochastic Effects. Especially with small doses of ionizing radiation, we cannot predict with any degree of certainty which individuals will show an effect. However, if a population is exposed, a certain fraction of the individuals will experience some effect, and this will show a higher incidence than in an unexposed population. Such an effect that can be detected only by statistical methods is called *stochastic*. An example would be the risk estimate of breast cancer induction by mammography.

Nonstochastic Effects. These appear predictably in any individual exposed to larger doses of ionizing radiation, such as the

160

development of a skin reaction in radiotherapy. For this reason, a nonstochastic effect may be called a *certainty effect*.

DOSE-RESPONSE MODELS

The severity or frequency of a particular effect of ionizing radiation depends on dose; that is, there exists a *dose-effect relationship*. However, the exact nature of this relationship is not yet fully understood, especially in view of the often inadequate or conflicting data. Consequently, several types of dose-response models have been proposed to help describe various radiation effects, although authorities disagree as to the applicability of a particular model to a specific effect.

The three most useful models, as expressed by their respective curves, include *sigmoid*, *linear*, and *linear-quadratic*.

Sigmoid Dose-Response Curve

Characterized by an S-shape, the sigmoid dose-response curve (see Figure 14.01) applies to high-dose effects prevailing in radiotherapy (Braestrup and Vikterlöf, 1974). Such a curve has a *threshold*; that is, below a particular minimum dose under given conditions, no observable effect occurs in the exposed individual. Above this minimum, predictable effects take place. Naturally, a different minimum dose may govern various effects. The S-shape means that a nonlinear relationship exists between effect and dose; the frequency or intensity of an effect is not proportional to dose. Furthermore, the shape of the curve at low doses denotes partial recovery from low doses. The curve eventually levels off and even turns downward at high doses because the animal or tissue dies before the effect appears.

The sigmoid dose-response curve has the following characteristics, according to Braestrup and Vikterlöf:

1. Presence of a threshold.
2. Partial recovery from smaller doses.
3. Dose-rate effect (decreased response at low dose rate).

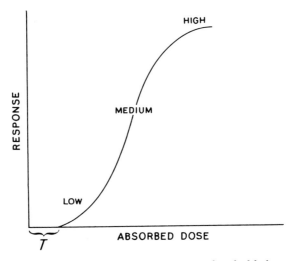

Figure 14.01. Sigmoid dose-response curve. T = threshold dose. As the dose increases, the response is not proportional but rather increases disproportionately for equal increments of dosage. At high doses, the effect levels off for equal increments of dosage because animals die before the specific effect can appear. Response is proportional to dose in the medium dosage range.

4. Plateau in response at upper limit of dosage; in fact, at very high dose levels curve may eventually turn downward.
5. Nonstochastic or certainty effect; predictable in the exposed individual.

We find the sigmoid-type response in radiotherapy wherein we can be reasonably certain that, for example, at a particular dosage level the skin will become red (erythema dose); or at another dosage level a severe mucosal reaction will ensue, as in the mouth; or at another dosage level a cataract will be produced. Therefore this type of response has been designated as a *nonstochastic* or *certainty effect.*

Linear Dose-Response Curve

In the linear dose-effect relationship there is *no threshold* (see Figure 14.02). Any dose, no matter how small, engenders an effect. However, an effect may be present even at zero dosage if there is a *natural* incidence of the same effect. Radiation simply increases the incidence with which it occurs. The linear dose-response curve represents a direct proportion between dose and response (i.e., frequency of occurrence or severity).

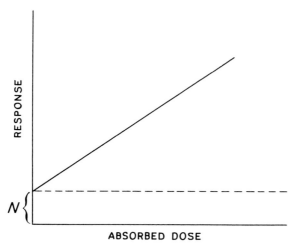

ABSORBED DOSE

Figure 14.02. Linear nonthreshold dose-response curve. There is no minimal effective dose—any dose, no matter how small, can induce the particular effect. At any dose level, the effect is proportional to the dose. The linear dose-response model is applied, for example, in estimating the upper limit of risk of radiation-induced genetic damage, leukemia, and breast cancer at low doses. *N* indicates the natural incidence of the particular effect in the absence of administered radiation.

We may summarize the characteristics of the linear dose-response relationship as follows:

1. No threshold.
2. Response (frequency or severity) proportional to dose.
3. No dose-rate effect (no reduced effect at small dose rates).
4. Stochastic (statistical) response; not predictable in any one exposed individual.

The linear dose-response curve applies to late effects that may or may not appear in a particular individual, but rather exhibit a statistically higher frequency in a population of individuals who have been exposed, in comparison with a population of like individuals who have not been exposed. In other words, such effects are *stochastic*.

One application of the linear dose-response model has been in estimating the genetic risk of ionizing radiation. Another has been in assessing the highest possible risk of inducing leukemia or breast cancer by low doses.

Linear-Quadratic Dose-Response Curve

In the 1980 report of the Committee on the Biologic Effects of Ionizing Radiation (BEIR), the majority proposed a linear-quadratic model for certain harmful effects of ionizing radiation such as cancer induction. The applicable curve (see Figure 14.03) is linear

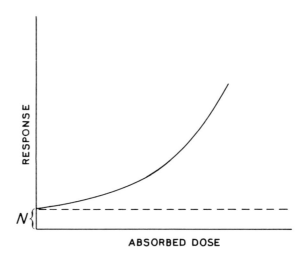

Figure 14.03. Linear-quadratic nonthreshold dose-response curve. At lower doses the response is linear, whereas at larger doses the response becomes nonlinear, that is, proportional to the square of the dose. This dose-response model has been recommended by the BEIR committee for estimating the carcinogenic risk at low doses. N indicates the natural incidence of the effect under consideration.

(proportional) at low dose levels and becomes curvilinear at higher doses. Thus, the upper portion of the curve fits an equation having a squared term (recall that a quadratic equation typically contains an x^2). A linear-quadratic dose-response effect has no threshold and is stochastic, just as is the purely linear situation.

However, they differ sharply in that the linear-quadratic model underestimates the effect at low doses. As we shall see later, when dose-effect data at high doses are extrapolated (extended) to the low-dose region, smaller (less numerous or less intense) effects are predicted than with the purely linear type response. This has led some authorities to opt for the linear model so as to err on the side of safety.

The characteristics of the linear-quadratic dose-response model may be summarized as follows:

1. No threshold.
2. Linear response at low dose levels.
3. Quadratic response at high dose levels.
4. Stochastic (statistical effect).

With this general background information, let us now turn to the late effects of ionizing radiation in human populations.

LATE SOMATIC EFFECTS

High-Dose Region

As stated earlier, we may arbitrarily regard doses of a few hundred rads or more as the high dose region. Much of the available information has come from animal experiments and from localized exposure in humans. The three main harmful effects at high dose levels include carcinogenesis, cataractogenesis, and life shortening.

Carcinogenesis (Cancer Induction)

Within just a few years after the discovery of x rays, they were found to produce skin cancer (in a radiologist in 1902). The induction of cancer by any agent such as radiation or certain chemicals is called *carcinogenesis*, and the inducing agent a *carcinogen*. In addition, certain substances may be needed to promote the action of carcinogens — these are known as *co-carcinogens* or *promoters*.

Radiation carcinogenesis in experimental animals is well known and unquestioned. In fact, it is dose related; Upton (1961) has shown that in mice subjected to whole body radiation, as the dose increases so does the incidence of leukemia (cancer of white blood cells) in the exposed population. However, at a dosage level of about 300 rads, the response reverses (see Figure 14.04) because many animals die from other radiation effects before developing leukemia. Such a response is not unique; a similar curve applies to other malignant tumors in various animals and there is no reason to suppose that humans would respond differently, although the dosage range may not be the same (i.e., dose-risk factor different).

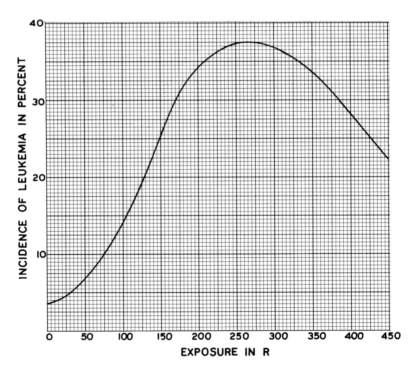

Figure 14.04. Frequency of leukemia in a mouse population receiving various x-ray exposures to the whole body. This follows a sigmoid dose-response curve. The curve turns downward at high dosage levels because increasing numbers of animals die before leukemia becomes manifest. (*After Upton AC, 1961*)

What evidence do we have for radiation carcinogenesis in humans? As we have just seen, this is manifested statistically by an increased frequency in an exposed population; that is, being a random process, the induction of cancer cannot be predicted in a particular individual. Radiation carcinogenesis involves long latent periods ranging up to 20 or 30 years for solid tumors. The following examples include some of the known radiation-induced cancers.

Breast Cancer. Induction of breast cancer by large doses of radiation has been well documented in humans, but experimental attempts to do this in small animals have been few in number. As pointed out by the NCRP in Report No. 66, the results in animals are confused by the effects of hormones and species differences.

The principal studies on radiation-induced breast cancer in humans include (1) Life Span Study, (2) Massachusetts Study, (3) Rochester Study, and (4) Nova Scotia Study. Although several others have been reported, these three have provided the most reliable data, and they will now be discussed briefly. Table 14.01 is a reanalysis of the data in the three series by the NCRP in Report No. 66 for the purpose of intercomparison.

LIFE SPAN STUDY. In this study of breast cancer induction resulting from the atomic bombing of the two Japanese cities (Hiroshima and Nagasaki) by the United States at the end of World War II, the exposed women have been followed since 1950 by various agencies, the last being the Radiation Effects Research Foundation (1980). Of the approximately 62,000 women in this study, one-fourth received a dose of about 10 rads and 108 of these eventually developed a breast carcinoma, an incidence of 0.70 percent. Of the remainder who received less than 10 rads, or no exposure, 243 incurred a breast carcinoma, an incidence of 0.5 percent. These results are subject to a number of uncertainties such as the size of the dose, the effects of heat and blast, the effect of whole body irradiation (including the ovaries, with possible attendant hormonal effects as in experimental animals), the loss of subjects through emigration, the 3 to 13 percent neutron component at Hiroshima (main type of radiation was gamma rays), and the possibly different risk levels in Japanese and Caucasian women. Table 14.01 summarizes the data from this study. Note the trend toward increased incidence of breast cancer with increasing dose, but no well-defined trend with age.

MASSACHUSETTS STUDY (Boice and Monson, 1976.) This is

TABLE 14.01

EXCESS NUMBER OF BREAST CANCERS
PER THOUSAND WOMEN PER YEAR
FOR EACH RADIATION STUDY, BY DOSE
AND BY AGE AT FIRST EXPOSURE.*

Series	Age at Exposure (years)	Breast Dose		
		0 rad	1-100 rads	100+ rads
A-bomb survivors (Tokugana et al, 1979)	10–19	0.17	0.27	0.87
	20–29	0.28	0.32	0.91
	30–39	0.34	0.28	1.60
	40–49	0.33	0.35	0.14
	50+	0.32	0.34	0.86
	For Series	*0.28*	*0.31*	*0.88*
Rochester mastitis patients (Shore et al, 1977)	15–19	0.00	0.00	4.08
	20–29	1.09	0.00	2.50
	30–39	2.53	3.45	4.31
	40–44	2.53	0.00	13.00
	For Series	*1.55*	*1.05*	*3.30*
Massachusetts fluoroscopy series (Boice and Monson, (Boice 1977)	10–19	0.53	1.36	2.36
	20–29	1.48	0.96	2.22
	30–39	1.09	1.66	1.58
	40–45	1.49	0.00	3.50
	For Series	*0.93*	*1.14*	*2.21*

*Adapted from *NCRP Report No. 66, Mammography*. This is an NCRP reanalysis with the three series in parallel.

The first five years of follow-up have been excluded.

based on the association between breast cancer incidence and repeated fluoroscopies of the chest in tuberculous women during induction and follow-up of therapeutic pneumothorax. It covered the period from 1930 to 1954. In 1047 women exposed to an average of 102 chest fluoroscopies, 41 breast cancers appeared (incidence 4 percent), as opposed to 23 expected breast cancers as the "natural" incidence. In the control group of 717 women with tuberculosis who did not receive fluoroscopy, 15 later developed a breast cancer (incidence 2 percent), compared with a "natural" incidence of 14 breast cancers. The average breast exposure was estimated to have delivered about 1.5 rad, with an average cumulative dose of 150 rads to each breast.

ROCHESTER STUDY (Shore et al., 1977). In this study 36 cases of breast cancer appeared in 571 women treated with x rays for acute postpartum mastitis at five or more years after irradiation, an incidence of 6.3 percent. Only 32 breast cancers occurred in 993 controls—patients with acute postpartum mastitis who did not receive x-ray therapy; in this group the incidence was 3 percent. Radiation doses averaged 377 rads per breast, given in one to 11 exposures. However, there is some question about a possible radiosensitizing effect of acute mastitis.

NOVA SCOTIA STUDY (Myrden and Hiltz, 1969). In a study of tuberculous women who had undergone repeated fluoroscopies for pneumothorax induction in Nova Scotia from 1940 to 1949, no dose estimate was given (technic stated), but the data strongly suggest radiation induction of breast carcinoma. Of 300 women subjected to multiple fluoroscopies and followed for 15 to 25 years, 7.3 percent incurred breast cancer, while in 483 controls the incidence of breast cancer was only 0.83 percent. Furthermore, there was a strong correlation between the side fluoroscoped and the incidence of carcinoma in the corresponding breast. Age at irradiation seemed to play a role, in that no increased incidence of breast cancer was observed in women who had received their fluoroscopies after the age of 30.

Table 14.01 summarizes the data from the first three studies. Note the absence of a real trend related to age of exposure. When all three studies are combined, there appears to be an increased susceptibility to breast cancer induction at ages 10 to 19 years (see

NCRP Report No. 66, page 41). The relation of breast cancer incidence to increasing dose is clearly shown in each study group.

Skin Cancer. Skin cancer occurred not infrequently in radiologists, orthopedists, dermatologists, and physicists during the early years of radiology. Some were caused by holding patients for immobilization in radiography. In too many instances the radiologist would demonstrate his hand to the patient during fluoroscopy. Individuals handling radium needles and other applicators without the use of long forceps also incurred a higher incidence of skin cancer.

Bone Cancer (Sarcoma). Occurrence of bone cancer was a nonstochastic effect in radium dial workers. In the 1900s women painted radium salt solution on the numbers and hands of timepieces to make them glow in the dark. These workers used their lips to shape fine tips on the brushes. Unfortunately they ingested enough radium which, as a bone-seeking element like calcium, selectively accumulated in the bones to cause osteogenic sarcoma years later. In fact, more than a thousand such cases have been reported.

Thorotrast. Years ago, thorotrast was used as a contrast medium in radiography of the liver and in angiography. These patients subsequently experienced an increased incidence of liver cancer, engendered by the radionuclide thorium in this contrast agent.

Lung Cancer. This occurs more often in uranium miners than in the general population. Inhalation of radioactive dust has been incriminated because the effect is dose related and we know that radiation is carcinogenic. Most miners who develop lung cancer have worked in the mines for 15 to 20 years, with a dose estimate ranging from about 0.01 to 1 rad per 40-hour week. Uranium miners in various parts of the world have an *excess* incidence of about 22 to 45 lung cancers per million persons per year per rad of alpha radiation to the bronchial epithelium (BEIR Report, 1980, page 391). Since the excess incidence of lung cancer in the atom bomb survivors (Japan) is about three per million persons per year per rad (mainly from gamma rays), the RBE for alpha particles is estimated at about 8 to 15. The effect of smoking on the incidence of alpha-induced lung cancer is probably additive rather than synergistic (i.e., multiplicative).

Ankylosing Spondylitis. Ankylosing spondylitis (rheumatoid arthritis of the vertebral column) used to be treated with x rays in some centers, the doses ranging from 400 to 2800 rads with suitable fractionation. A British report showed that some years later the treated individuals experienced a twelve-fold increase in the occurrence rate of leukemia (Court-Brown and Doll, 1965). About 80 percent were the acute myelogenous type. Despite some question as to the validity of the initial conclusions, they were later substantiated, so that the radiation-induced origin of leukemia in these patients has now been generally accepted. Noteworthy are the strong dose-effect relation (increasing incidence with increasing dose) and the predominance of myelogenous leukemia, the same type that has occurred most frequently in the Japanese survivors of the atomic bomb.

Thyroid Carcinoma. A higher than expected incidence of thyroid carcinoma has been found in persons who have undergone irradiation therapy of a so-called enlarged thymus during infancy and childhood (Hempelmann, 1968; Hempelmann et al., 1975). Dose estimates have ranged from about 50 to 300 rads (0.5 to 3 Gy). Irradiation of patients with facial acne and scalp ringworm has also been associated with an increased frequency of thyroid carcinoma. According to the BEIR Report (1980), the estimated *excess* incidence of thyroid cancer induced by radiation is about four cases per million persons per year per rad. These radiation-induced thyroid cancers are most often papillary carcinomas of such low grade that they do not add significantly to mortality. The average latent period between exposure and appearance of cancer is about 10 years.

Atom Bomb Survivors. In the Hiroshima and Nagasaki (Japan) atom bombings, survivors have experienced an increased incidence of leukemia and other cancers as well (see above, Breast Cancer). Acute and chronic myelogenous leukemias predominate. The increase has become apparent at one and one-half to five years after exposure, reaching a peak at 14 years, and then declining. After 20 years the incidence approximates the natural incidence of leukemia in the same population. In Hiroshima the leukemia excess occurred in persons located within 1500 meters from the hypocenter (site at which bomb struck), but for unknown reasons persons in Nagasaki beyond 2000 meters from hypocenter still showed an excess of leukemia. In general, the risk of leukemia rose above the natural incidence at a rate of one or two cases per million population per year per rem, averaged over the 14 years included in the study of Brill and others (1962).

Cataractogenesis

A cataract is a localized or generalized loss of transparency in the lens of the eye. Although cataracts appear frequently with advancing age, they can also occur after irradiation. In the earlier stages, a radiation-induced cataract differs from a natural cataract in that the former starts in the posterior central portion of the lens. If it fails to progress it may become doughnut shaped with a clear center and an opaque periphery. Otherwise, complete opacification of the posterior pole of the lens takes place.

What is the mechanism of radiation cataractogenesis? Under normal conditions the epithelial cells in the front part of the lens produce transparent fibers throughout life (vertical renewal system). Following irradiation the damaged cells make abnormal fibers, which accumulate in the posterior pole of the lens and form opaque spots that make up the cataract. Some animals such as mice are highly susceptible to cataractogenesis, requiring only a few rads of low-LET radiation. Humans show less sensitivity in this regard; the minimum dose of low-LET radiation known to induce a human cataract is 200 rads (2 Gy), but this is nonprogressive. At 500 rads (5 Gy), the threshold for progressive cataract, complete opacification eventually occurs.

The latent period in humans varies inversely with dosage and ranges from about six months to 35 years. According to Hall (1978) the average latent period following a dose of 250 to 650 rads (2.5 to 6.5 Gy) of low-LET radiation is about eight years, decreasing to about four years with doses of 650 to 1150 rads (6.5 to 11.5 Gy).

High-LET radiation, particularly neutrons, is especially effective in cataractogenesis. In fact, the maximum RBE of fast neutrons is about ten for doses of a few hundred rads. This means that the minimal cataractogenic dose of neutrons is one-tenth that of x rays, or about 20 rads. At very low doses of a few rads the RBE for fast neutrons rises to about 50 because x rays are much less

cataractogenic at low doses. With suffciently large doses of ionizing radiation, cataractogenesis becomes a certainty effect.

Life Shortening

Except for large doses to the whole body, such as those responsible for the acute radiation syndromes, life shortening is a stochastic
effect and is nonspecific in character. Animal experiments have
shown a drop in average survival of animals in colonies exposed to
radiation, the outcome being dose-dependent — the greater the
dose, the shorter the life expectancy of the exposed individuals.
Note that we cannot predict which animal will experience life
shortening; we are dealing with statistical averages.

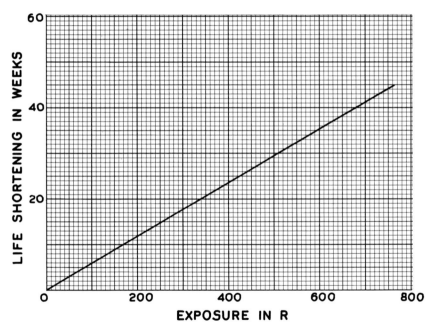

Figure 14.05. Life shortening in a mouse population receiving various x-ray
exposures to the whole body. Note the linear relationship between dose and effect,
at least in this range. (*Adapted from Rotblat J, Lindop P. Proc R Soc [London] 154:350,
1961.*)

Figure 14.05 is a graph based on the work of Rotblat and Lindop
(1961) who exposed mice to various doses of x rays ranging from 50

to 800 rads (0.5 to 8 Gy). It shows a life span reduction of about 5 percent for each additional 100 rads (1 Gy) single whole body dose. Postmortem examination revealed only a speeding up of the aging process: a reduced number of parenchymal cells and blood vessels, and an increased amount and density of connective tissue throughout the body. These changes may actually have resulted from destruction of stem cells and eventual loss of differentiated cells through nonreplacement, accompanied by degeneration of connective tissues and blood vessels. Besides the nonspecific, premature aging through degenerative processes, shortened life expectancy results from an increased rate of carcinogenesis.

Insofar as *humans* are concerned, we have no firm evidence of radiation-induced life shortening. Statistical studies on the longevity of atom bomb survivors (other than excess mortality from cancer) as well as the most recent ones on the longevity of radiologists have failed to show conclusively any reduction in life span.

Low-Dose Region

Radiodiagnostic procedures, including x rays and radiopharmaceuticals, involve low-LET radiation in low doses. We should therefore expect to find somatic and genetic *stochastic* effects because these typically result from such exposure and appear at random. Thus, we face the problem of estimating the risks incurred in radiodiagnosis. We must constantly balance the benefit of a particular x-ray or radionuclide procedure against the risk of inducing a harmful effect that may overshadow any anticipated benefit. We call this relationship the *benefit/risk ratio*.

Three main kinds of low-dose effects predominate. Two are *somatic*: carcinogenesis and injury to embryo and fetus. The third is *genetic*.

Carcinogenesis

We have already discussed carcinogenesis as a high-dose effect. Carcinogenesis is also the *major* late somatic effect of low-dose low-LET radiation. Myelogenous leukemia predominates as a detectible radiation-induced cancer mainly because of its small natural incidence and its short latent period—two to four years from exposure.

However, the total increase in incidence of *all* solid tumors resulting from low-dose irradiation exceeds that of leukemia.

Estimating the excess number of cancers induced by low-dose radiation has thus far been subject to a large uncertainty factor. Unfortunately, we have no simple or positive way to assess benefit/risk ratio. We must assume at present that if a certain, large dose of radiation causes a particular incidence of a certain effect, a small dose will produce a smaller incidence of the same effect. However, this relationship depends more on the shape of the selected dose-response curve than on the available data. *The question is how best to extrapolate high-dose data to the low-dose region.*

In the 1980 report of the Committee on the Biologic Effects of Ionizing Radiation (BEIR) of the National Research Council (supported by the Environmental Protection Agency), a great deal of attention was paid to the choice of the most appropriate dose-response model for the low-dose region. While the Committee agreed that, in the absence of evidence to the contrary, we must assume a nonthreshold type response (see pages 162–165), with a curve that is linear (straight line) at low dose levels, it was not unanimous as to the shape of the curve at upper dosage levels. In this BEIR Report (1980) the Committee struck a compromise by assuming a linear-quadratic (LQ) curve, which is straight at low dose levels and merges with a quadratic (upward curving) curve at higher doses, as already described on pages 164–165. Such a dose-response curve has been shown in its general form in Figure 14.03. It should be emphasized that the purely linear nonthreshold dose-response at low doses errs on the side of safety (overestimates risk) which is important in radiodiagnosis. Furthermore, there is experimental evidence for partial recovery at low doses with low-LET radiation (but not with high-LET radiation).

We shall now discuss two types of cancer that may be induced by low-dose irradiation: leukemia and breast carcinoma.

Leukemia. More evidence exists for the induction of leukemia (cancer of white blood cells) by low-dose irradiation than for any other type of cancer. A number of retrospective (review of patients' histories) and prospective (future incidence) studies on low-dose radiation-induced leukemogenesis have been reported, and we shall now summarize the more important ones.

Court-Brown *et al* (1960) conducted a prospective study of

nearly 40,000 live births and found no increased incidence of leukemia in children whose mothers had received x-ray examinations involving the pelvis and abdomen during pregnancy.

On the other hand, MacMahon (1962) reported a prospective study of nearly 750,000 children in northeastern United States, irradiated *in utero* during radiography of the mother's pelvis or abdomen. They found an overall increase in general cancer incidence by a factor of 1.42. This was virtually the same for leukemia, cancer of the central nervous system, and other types of cancer.

Although there is still some question as to the carcinogenic potential of such small doses of ionizing radiation, most authorities accept a risk factor of 1.40 according to present evidence, incompletely substantiated as it may be. Note that this applies to the fetus that has received diagnostic x rays with a fetal dose of perhaps 400–600 mrems.

According to the latest BEIR Report (1980), the *lifetime risk* in terms of *excess* cases of leukemia (and bone cancer) from a single 10-rad dose of x rays to the whole body is about 230 per million population, or 2.6 percent above the anticipated "natural" incidence. For an *annual* dose of one rad over a lifetime, the excess incidence is about six times that for a single 10-rad dose (see Table 14.02). These data are based on the LQ-L model in which extrapolation from the high-dose to the low-dose region follows a linear-quadratic curve for gamma rays and a purely linear curve for neutrons. The RBE for neutron induction of leukemia was taken as 11.3

With an L-L model, (linear for both gamma rays and neutrons), which probably represents the upper limit of risk, the excess incidence is about twice as great or 500 per million population; many authorities would prefer this risk estimate over the one obtained with the linear-quadratic model.

You may ask, "What is the risk of a single whole-body dose of one rad?" We may safely assume on the basis of present knowledge that it would be one-tenth that from a single dose of 10 rads. Thus, *one rad would probably entail a lifetime excess of about 50 cases of leukemia (and bone cancer) per million population*, based on a purely linear dose-response curve. The delivery of one rad to a fetus could easily be reached during excretory urography of the mother

TABLE 14.02

LIFETIME EXCESS INCIDENCE OF LEUKEMIA AND BONE CANCER (*BEIR 1980*) PER MILLION EXPOSED POPULATION.**

L-L MODEL***—LOW–LET

Dose (whole body)	Expected Incidence per Million	Increase per Million	%Increase
10 rads, single dose	8940	475	5.3
1 rad/yr, lifetime	9800	3140	32.0

LQ-L MODEL****—LOW–LET

Dose (whole body)	Expected Incidence per Million	Increase per Million	%Increase
10 rads, single dose	8940	230	2.6
1 rad/yr, lifetime	9825	1400	14.02

**From BEIR Report (1980), based on Tables V-16 and V-17. Present author has averaged and rounded off the data for male and female populations; risk estimates slightly higher for men than for women.

***Linear for both gamma rays and neutrons.

****Linear-quadratic for gamma rays, linear for neutrons.

Note that with the L-L model, the risk is about twice that with the LQ-L model. In the original data, with progressive increase in the age at which a continuous lifetime exposure of one rad/year was started, the percent increase in excess incidence decreases.

using only four abdominal radiographs (par speed screens).

Breast Cancer. We are concerned here with the implications for breast cancer induction resulting from mammography. This obviously involves mammographic doses to the breast tissue ranging downward from one rad. As noted earlier (see pages 167–168) we have no direct evidence for induction of breast cancer in the low-dose and low dose-rate region with low-LET radiation such as x rays. Risk estimates for low doses depend on extrapolation (extension) of dose response curves from high-dose to low-dose levels. The paramount question here is the shape of the curve in the low-dose region. The NCRP Report No. 66 recommends a *purely linear nonthreshold dose-response model* to determine the probable *upper limit of risk for mammography.*

Accordingly, we assume that the cancer risk *per rad* does not diminish with decreasing dose. The minimum latent period from the time of exposure to the first appearance of cancer is about five years in women age 30 or more, but another five years must elapse before a substantial increase in risk becomes manifest. The size of the dose has no effect on the length of the latent period. Because the increased risk persists for 30 to 45 years, we must assume a lifetime risk from any exposure, no matter how small. Finally, maximum risk probably exists when exposure occurs in the 10 to 19 year age group.

Two models have been proposed for specifying risk estimates:

1. *Absolute risk* is the number of cancers over and above the number occurring spontaneously (i.e., without irradiation), projected forward in time from the age at exposure. Thus, absolute risk is the number of cancers in the study group minus the number in the control (unexposed) group. Absolute risk is usually stated in terms of the excess cancers per million exposed women per year of observation per rad.

2. *Relative risk* is simply the ratio of the observed number of cancers in the study group to the number in the unexposed group. That is, it is the number of observed cancers divided by the number expected. Inherent in this model is the assumption that risk depends also on the age at observation (and not only at age of exposure, as in the absolute risk model).

There is no consensus as to which of these two risk models is to be preferred, and so both are usually included in published reports. Table 14.03 gives the range of estimated absolute risk for human breast carcinoma, as adapted from BEIR Report (1980) according to the data of Land. Note the absence of trend in incidence with age, except for the higher risk in ages 10 to 19.

At present, based on such data extrapolated along a purely linear dose-response curve with no threshold, the risk estimates for human breast carcinoma induction by x rays at low doses are:

absolute risk = 6.6 ± 1.9 (approx 6) excess breast cancers per million women per year per rad.

relative risk = 0.53% ± 0.15 (approx 0.5%) increase in relative risk per rad.

TABLE 14.03

RISK ESTIMATES FOR BREAST CARCINOMA.*

| | | LINEAR MODEL—LOW-LET | |
Series	Age at First Exposure	No. of Breast Cancers in Women Exposed to 1 Rad	Range of Estimated Absolute Risk* per Rad
atom bomb survivors	10–19	40	7–11
	20–29	36	2–4
	30–39	28	2–8
	40–49	15	—
	50+	15	1–6

*Excess cases per million exposed women per rad per year of life after 10 years after exposure, or age 30, whichever is later. (*Adapted from Land, CE et al as presented in BEIR Report, 1980, Table A-3, p 280.*)

Note greater radiosensitivity of breasts for carcinoma in younger women; similar trend in other studies such as repeated fluoroscopies in tuberculous women, and in those receiving radiotherapy for mastitis.

Aside from the selection of the most suitable dose-response curve for extrapolation from high doses to low doses, *statistically valid* risk estimates from small doses in humans require large populations in terms of number of subjects and number of years to observe the full impact of the risk. The population sample (i.e., the number of subjects times the number of years' observation) is inversely proportional to the square of the dose. Thus, as the test dose is decreased, there is a rapid rise in the required population size. For example, as indicated in NCRP Report No. 66, with a dose of 10 rads in a single dose to the breast we would need a population of 250,000 women in the exposed group, and an equal number in the control (unexposed) group, all aged 50 years. If they all lived 20 years, 5700 cancers would occur spontaneously (i.e., natural occurrence rate) after age 60 (minimal latent period 10 years). If in the exposed group we assume an excess risk of 6 cancers per million women per year per rad, there would be 150 cancers presumably induced by the 10 rads in ten years after age 60. (6 ca/million = 1.5 ca/250,000. 1.5 ca × 10 rads × 10 yrs = 150 ca.) An excess of 150 cancers above the "natural" incidence of 5700 cancers would have borderline statistical validity. However, if the exposed women were to receive a breast dose of only one rad, this being in the mammographic range, there would be only 15 excess breast cancers, but in this case 25 million women would be required in each group to obtain a statistically valid risk estimate. In other words, unless one were to study an inordinately large population, the estimate of risk would be subject to a very large error.

In mammography with presently available technic and imaging systems, the dose should be less than one rad at the breast surface for each exposure. Thus, with an average depth dose of 50 percent in the center of the breast (depending, of course, on the thickness and consistency of the breast) the midplane breast dose should be less than one rad for a two-view examination. According to Jans et al. (1979), 63 percent of mammograms are performed by xerography, and 35 percent by film-screen low-dose technic. He has reported the midplane doses for a two-film examination with these two methods. The doses have been averaged as follows:

Xeromammography	0.75 rad
Film-screen low-dose	0.15 rad

There has been considerable debate about the mass screening of asymptomatic women for breast cancer. One question pertains to the possibility that as many breast cancers might be induced by mammography as might be detected (Bailar, 1976). Possible answers to this question lie in two screening studies. In the Hospital Insurance Plan of New York (HIP) reported by Strax *et al.* (1973) among about 62,000 women in age range 40 to 60 years, 299 breast cancers were detected by film mammography within five years. Of these, 33 percent were discovered by mammography alone, of which 13 percent were classed as minimal.

In the other screening study, the 27 Breast Cancer Detection Demonstration Projects (BCDDP) sponsored by the National Cancer Institute and the American Cancer Society (Beahrs *et al.*, 1979), of 280,000 women examined by physical examination, mammography, and thermography, 2,862 cancers were detected—45 percent by mammography alone. It is noteworthy that 40 percent of the cancers were less than one cm in diameter and that nearly one-half the women in the surveys were below age 50; 30 percent of the cancers occurred in this younger age group.

Since, as we have noted before, the upper limit of risk for breast cancer from low doses of ionizing radiation is about 6.6 cancers/million women/year/rad (lifetime), the risk for a two-view mammogram with a surface dose of one rad per view, a depth dose of 50 percent, and a 20-year period after exposure would be as follows:

$$2 \; rads \times 0.50 \times 20 \; yr \times 6.6/10^6/yr/rad =$$
$$132 \; cancers/10^6 \; women \; screened$$

Now consider the benefit of mammography. As summarized by Feig (1977) 45 percent of cancers were detected by mammography alone in women aged 35 to 65 years, or 3.6 cancers per 1000, representing an overall detection rate of eight cancers per thousand women. Of these occult cancers, 90 percent had negative axillary lymph nodes (85 percent being less than 2 cm in diameter). In these favorable circumstances, we should expect a 90 percent five-year survival rate. Therefore, the expected benefit in terms of lives saved by mammography would be:

$$benefit = 3.6/1000 \times 0.9 \times 0.9 = 2.9/1000 =$$
$$2900/10^6 \; women \; screened$$

In summary, the upper limit of risk may be 132 cancers per million women if a skin dose of one rad was delivered for each of two mammographic exposures. The benefit may be 2900 lives saved. However, you must realize that these are simply estimates based on the best available, although not entirely reliable, data.

Table 14.04 gives the estimated benefit/risk ratio for xerography and low-dose screen-film mammography with the best available technic. By either method, there is a large benefit/risk ratio; that is, many more cancers would be detected than might be induced, even assuming a linear, nonthreshold dose-response curve at low doses. In actual practice, the risk may be even smaller than indicated because the linear model may overestimate the risk.

Despite the detection of a significant number of cancers by mammographic screening of asymptomatic women between ages 35 and 50, the NRCP does not recommend screening below age 50. However, if there is clinical suspicion of cancer or if the individ-

TABLE 14.04

GROSS ESTIMATES OF BENEFIT/RISK RATIO FOR
MAMMOGRAPHIC SCREENING OF WOMEN AGE 35 AND OLDER.
RISK OF CANCER BASED ON LINEAR NONTHRESHOLD
DOSE–RESPONSE CURVE.*

Method	Midplane Dose Two Views (rads)	Possible Cancers Induced in 10^6 Women†	Benefit Lives Saved in 10^6 Women	Benefit/Risk Ratio
Xerography	0.75	99	2900	2900/99 = 29/1
Film-screen				
(low-dose)	0.15	20	2900	2900/20 = 145/1

*Modified from Feig, In *Breast Carcinoma*, 1977.
†Lifetime excess incidence (midplane doses × 20 years × 6.6 ca/10^6 women/yr).

ual is in a high cancer-risk category, mammography is not contra-
indicated below age 50.

Other Types of Cancer. Fatal cancers other than those involving
the hematopoietic system (leukemia), bone, and breast may be
induced by low-dose low-LET radiation. These are listed in Table
14.05, abstracted from the BEIR Report (1980). Note that, with the
exception of leukemia and breast cancer, the solid cancers of the
thyroid, lung, and stomach lead the list of radiation-induced
malignancies in men and women collectively. The data in this
table derive mainly from atom bomb survivors (Japan) and have
been extrapolated to low-dose levels from high-dose levels according to a modified linear-quadratic type of dose-response curve.
For all sites the average excess risk is 13 cases per million males
per year per rad, 11 to 30 years after exposure. The risk is about
twice this (actually, 23) in females. Furthermore, using a modified
linear dose response curve, the BEIR reports virtually twice as
great a risk as that just quoted.

Radiation Injury to Embryo and Fetus

We found earlier (pages 152–156) that doses at the 200-rad level
cause profound changes in mammalian (mouse) embryos, particularly during the organogenesis stage which, in humans, occurs
from the fourth to the eleventh week of gestation (before 2 or 3
weeks, embryonic death from this dose of radiation often takes

TABLE 14.05

ESTIMATED EXCESS CANCER INCIDENCE
(EXCLUDING LEUKEMIA AND BONE CANCER)
PER MILLION PERSONS PER YEAR PER RAD,
11 TO 30 YEARS AFTER EXPOSURE,
BY SITE, SEX, AND AGE AT EXPOSURE.*

| | Age at Exposure, yr | | | | | |
Site	0–9	10–19	20–34	35–39	50+	Age-weighted Average
MALES						
Thyroid	2.20	2.20	2.20	2.20	2.20	2.20
Lung	0.00	0.54	2.45	5.10	6.79	3.64
Stomach	0.40	0.40	0.77	1.27	3.35	1.53
All Sites	4.80	5.29	9.11	13.66	22.59	13.
FEMALES						
Thyroid	5.80	5.80	5.80	5.80	5.80	5.80
Breast	0.00	7.30	6.60	6.60	6.60	5.82
Lung	0.00	0.54	2.45	5.10	6.79	3.94
Stomach	0.40	0.40	0.77	1.27	3.35	1.68
All Sites	8.40	16.19	19.31	23.86	32.79	23.

*Data extracted from *BEIR Report (1980)*, Table V-14, p 198. These are strictly estimates based on various assumptions; the risk estimates are based on the linear-quadratic model, and would be higher (possibly twice as high) with a purely linear model. Risk estimates for organs with excess cancer risk less than 1.5 have been omitted, although present in the original table. The total for "all sites" is one possible measure of the effect (excluding leukemia and bone cancer) of whole body radiation with all tissues receiving one rad of low-LET radiation.

place, but very few abnormalities appear in embryos surviving to term). The question now arises as to the effect of *low doses* of ionizing radiation in the diagnostic region. Hammer-Jacobson (1959) has stated that a dose of 10 rads or more during the first six weeks of pregnancy is so likely to cause serious genetic damage that the pregnancy should be promptly terminated (i.e., medically indicated abortion). Although it has not been universally accepted, many authorities tend to follow this recommendation.

The severe radiation effects in early pregnancy have prompted the application of the so-called "ten-day rule," especially in England. However, some experts such as Gray (1979) have pointed out that if a potentially pregnant woman has to have an x-ray examination postponed, and if she turns out actually to be pregnant, the examination would have to be delayed for nine months; therefore he would question the need of the examination in the first place. On the other hand, if the examination cannot be postponed on account of the urgent medical reasons, then it should be performed, since the benefit would presumably outweigh the risk. Special precautions must be taken to minimize exposure of the embryo by careful collimation and reduction in the number of films when possible; by the use of fast film-screen technic; or by the use of other modalities such as ultrasound (although the risk factor here has not yet been firmly established).

In the event that a large diagnostic x-ray workup has been performed, some authorities advise a 10-month wait for the mother and a 3-month wait for the father before conception of the next child. These intervals enhance the probability that an *un*irradiated ovum and sperm will enter into the fertilization process.

GENETIC EFFECTS OF RADIATION

In Chapter V we found that ionizing radiation induces in reproductive cells changes that may be transmitted to future generations. We designate this the *genetic effect of radiation*. It was first reported by Bardeen (1907), who found that when he exposed toad sperm to x rays and then used the same sperm to fertilize toad ova, abnormalities appeared in the offspring. We have noted earlier (pages 66–67) that such genetic changes result from mutations (abrupt changes) in genes or chromosomes. In fact, H. J. Mueller (1927) conclusively demonstrated the induction of gene mutations in fruitflies (*Drosophila sp.*) by exposure to x rays, and was even able to trace the mutant genes through several generations.

Production of Gene Mutations and Chromosome Aberrations

Important as this subject is, we must depend almost entirely on animal experiments for our knowledge about the processes involved in the production of gene mutations and chromosome aberrations. The following outline aims to review the more important available data.

Gene Mutations

Each gene consists of hundreds of thousands of nucleotides in a characteristic sequence (see pages 46–49), that is, DNA segments. Such sequences consist of nucleotide triplets, each of which encodes a specific amino acid, although the process is not limited solely to protein manufacture (see pages 51–54). We shall now review the important characteristics of gene mutations.

Mutagenic Agents. These include not only ionizing radiation but also a host of chemical agents that can disturb the nucleotide sequence in DNA, or even delete or add nucleotides. Gene mutation involves the breaking of chemical bonds among the constituents of the DNA molecule. Although the vast majority of mutations in laboratory animals are harmful in one way or another, important exceptions occur such as the decreased susceptibility of insects to DDT, and of bacteria to antibiotics.

Radiation-Induced Mutations Recessive. A preponderance of radiation-induced mutations are *recessive.* Their effects on future offspring become manifest only if the individual harboring them happens to mate with another bearing the same mutant gene in the other member of the chromosome pair, or if the mutant gene is located in the X chromosome of a son of an irradiated mother (see pages 64–68).

Delay in Appearance of Mutation. The usually recessive nature of radiation-induced mutations dictates a delay in their appearance for a number of generations, but they may never show up at all. In fact, some mutations may be lethal, causing death of the embryo at such an early stage of development that we are unaware of their occurrence. A mutation in *one* X chromosome of a female, even though it may be recessive, will become manifest in about one-half of her male offspring in the next generation. (Why? See pages 44–45.)

Radiation-Induced Mutations Not Unique. There are no characteristic differences between radiation-induced and spontaneous mutations. Radiation merely increases the mutation frequency, that is, more mutations per generation. The effect is dose-dependent.

High-LET Radiation More Effective. For equal doses, high-LET radiation is more effective (up to a limit) than low-LET radiation in causing gene mutations.

Linear Dose-effect Curve. Down to a dosage level of a few rads the dose-effect curve is known to be linear (see pages 162–163), so it is generally assumed that *no threshold exists*; that is, any dose, no matter how small, can induce a gene mutation.

Dose-Rate Dependence. There is sufficient evidence at present that low dose rates (i.e., very small dose per second), as well as dose fractionation, are less likely to cause mutations than are high dose rates. This undoubtedly depends on repair mechanisms (Elkind or fast recovery) that occur during the irradiation process at low dose rates or with fractionation.

Mutagens, both radiation and chemical, can affect human health and still go unrecognized. In the first place, even with a high mutation rate induced by these agents, the resulting mutations may be so hidden by other factors (delayed appearance; coexistence of other abnormalities) that it is extremely difficult to blame them on any particular mutagen. In the second place, radiation and chemical mutagens do not engender specific mutations; they only increase the frequency of mutations that occur spontaneously.

Chromosome Aberrations

Loosely called chromosome mutations, these include principally *breaks* with subsequent loss or rearrangement of chromosome fragments, ranging from single nucleotides to large chromosome fragments (see pages 63–64). Chromosomes consist of genes lined up like beads on a string. We cannot always differentiate between a gene mutation and a chromosome aberration. With low-LET radiation (x or gamma rays) the rate of chromosome aberrations is nearly proportional to the dose in the dosage range that is of interest to us.

Naturally Occurring
Complex Inherited Disease in Humans

Much has been written about the incidence and nature of hereditary diseases in man. Since they have existed long before the modern era of radiology, we may assume that they result from spontaneous mutations. Based on the studies of McKusick (1978), complex inherited diseases in man may be classified as follows.

Gene Mutations in Humans

Dominant or X-linked Diseases. These occur in 1 percent of children, appearing in any individual harboring the mutant gene. There are about 415 known dominant or X-linked human genetic diseases caused by such spontaneous mutations, although what percentage is induced by radiation is unknown. Examples include achondroplasia, polydactyly, retinoblastoma, certain types of muscular dystrophy, and a particular type of renal carcinoma. The expressed incidence is approximately proportional to the mutation rate, that is, the higher the mutation rate the more likely is one of these diseases to occur.

Recessive Diseases. These appear in less than 0.5 percent of the total genetic disease incidence (not of the total number of children) and may not become manifest for many generations, or may even be lost through decreased fertility or viability. In fact, there is a good chance that such a recessive gene may be lost before it has had an opportunity to be expressed. There are probably 365 recessive genetic diseases, including, for example, cystic fibrosis, sickle cell anemia, PKU, and Tay-Sachs disease; others are extremely rare. A recessive gene on an X chromosome acts like a dominant in a son who carries it because the homologous Y chromosome carries no genes. Falling into this category are conditions such as hemophilia, color blindness, and a severe form of muscular dystrophy.

Anatomic Malformations. These occur in 0.5 percent of births.

Mixtures of Constitutional and Degenerative Diseases. Mixtures occur in about 1.5 percent of individuals on a genetic basis, but this remains an uncertain category. It includes, as examples, diabetes mellitus, anemia, schizophrenia, and idiopathic epilepsy, but not most cancers, heart disease, or peptic ulcer. However, a genetic predisposition may be present in some individuals for certain diseases, which require environmental factors for their expression.

Thus, the above groups comprise about *6 percent* of all live births. However, we do not as yet have an accurate method of estimating the total impact of gene mutations on humankind. This applies especially to very mild genetic changes that may be a long time in appearing or may not be recognized when they do appear.

Chromosome Aberrations in Humans

Changes in the number of chromosomes or fragmentation of chromosomes with or without subsequent rearrangement of fragments are known as chromosome aberrations.

Aneuploidy. In order to understand the meaning of aneuploidy we must first explain *euploidy.* As we have noted earlier (pages 36–37) one full set of chromosomes is called the *haploid* number; in humans this set consists of 23 different chromosomes. *Euploidy* is a condition in which a cell has one or more *complete* sets of chromosomes, so if a cell has two sets, as in normal somatic (body) cells, it is a *diploid* cell. If it has three sets it is a *triploid* cell and the condition is designated *triploidy*; and if four sets, *tetraploidy.* Triploidy or tetraploidy of all cells in the human body is incompatible with life; rarely, survival is possible in a so-called mosaic individual in whom some cells are diploid (normal) and others triploid.

Aneuploidy is a condition in which cells have an abnormal complement of chromosomes, that is, an excess or deficiency of certain chromosomes. An important example is Down's syndrome (previously mongoloidism) or trisomy 21, wherein there are three instead of two chromosomes designated as 21. Chromosomes are classified according to size and configuration—each pair is assigned an identification number as shown in Figure 14.06. Down's syndrome is caused by *nondisjunction* (failure to separate) of chromosomes 21 during meiosis of an ovum so that it retains two chromosomes 21. When this ovum is fertilized by a normal sperm the resulting individual has three chromosomes 21, hence the term trisomy 21. According to the 1980 report of BEIR (pages 92, 125), although nondisjunction and trisomy can be induced by ionizing radiation, the risk is small because it requires very large doses. More often, aneuploidy of other kinds leads to death of the embryo.

Double Chromosome Breaks. Double breaks are less common than aneuploidy. They may result in exchange of fragments between two chromosomes, a process called *translocation.* Obviously this involves a break in one or more chromosomes. If each chromosome ends up with its full complement of genes, we have a *balanced translocation* and the resulting individual is normal. If not, the

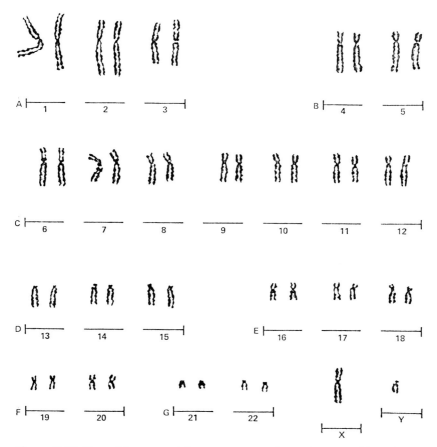

Figure 14.06. Normal karyotype (chromosome complement) prepared from human leukocytes dividing in culture. After suitable preparation, the chromosomes appear in a cluster as seen through a microscope. A photograph is made and the individual chromosome pictures are cut out so the chromosomes can be sorted in homologous pairs. These are finally numbered and pasted up according to the Denver classification as shown here. In this example, the presence of an *X* and a *Y* chromosome means that this is a male karyotype.

individual has an *unbalanced translocation* and may exhibit physical or mental abnormalities. The most common type of balanced translocation is the *Robertsonian* in which each chromosome fragment retains its centromere and spindle fiber with eventual fusion of the fragments and spindle fibers as shown in Figure 14.07. The incidence of Robertsonian translocation is eight out of 1000 live

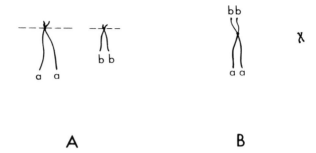

A   **B**

Figure 14.07. Robertsonian translocation. In *A*, a break has occurred at the centromeres of two different chromosome pairs, indicated by *1* and *2*. Note that there are two acrocentric chromosomes, with the centromere at one end.

In *B*, the long arms of the broken chromosomes join at their centromeres. The short arms are lost, but usually carry little or no significant genetic information. When the chromosome in *B* divides during subsequent mitosis, each daughter cell carries the full complement of genetic material, and so the cells are normal. However, gametes are not always normal and the resulting individual in a later generation may be abnormal. Robertsonian translocation does *not* often follow irradiation in humans.

births, such individuals usually appearing normal.

Single Chromosome Breaks. Single breaks may heal or give rise to complex inheritance if the fragments become maldistributed during meiosis. Examples include schizophrenia, mental retardation, and certain types of cancer.

Extremely Mild Chromosome Aberrations. Mild aberrations may lead to such conditions as reduced longevity, susceptibility to disease, slight anatomic malformations, or impaired vigor. Because of their mild effects, such aberrations tend to remain submerged in the genetic pool longer than severe aberrations, and therefore ultimately affect more people, although they may be almost undetectable.

Risk Estimates for Genetic Damage

Insofar as man is concerned, we have no direct proof for radiation-induced *gene* mutations. However, we can reasonably assume that the lesions produced in various animals are basically similar to those in man, although the relationship to certain factors such as dosage may not necessarily be the same. At the same time, we do have reliable evidence for the induction of chromosome aberrations by radiation in man.

For low doses and low dose rates of low-LET radiation, BEIR (1980) extrapolates from high-dose data in mice, using a linear dose-response curve. Such data are used to estimate the genetic risk of radiation in man.

We shall now consider the increased risk of radiation mutagenesis relative to the spontaneous mutation rate in humans, but first we must define two terms.

Genetically Significant Dose (GSD). As shown in Table 15.01, the two major sources of human exposure to low level radiation are medical and dental x rays, and natural background. As for medical and dental x rays, only a *part* of the population is exposed, and so the GSD has been introduced by the Bureau of Radiological Health as a mathematical statistical concept to evaluate the total genetic impact of such radiation. The *GSD is defined* as that gonadal dose of ionizing radiation which, if received by every individual of the population, would be expected to produce the same total genetic injury to the population's future offspring as would the total dose actually received by the various individuals. The GSD is also weighted for the number of offspring the exposed individuals would be likely to have. Thus,

$$GSD = \Sigma D_i \hat{N}_i P_i / \Sigma N_i P_i$$

where D_i = average gonad dose to persons of age i who receive x-ray examinations, $\hat{N}_i$ = number of persons in population of age i who receive x-ray examinations, P_i = expected future number of children of persons of age i, and N_i = total number of persons in population of age i.

On the other hand, for background radiation the GSD and the gonadal dose are the same because the entire population is continually subjected to low level exposure. In 1970 the GSD from diagnostic x rays was 20 mrems per year, of which $2/3$ involved males and $1/3$ females.

Mutation Rate or Frequency. The number of mutations, spontaneous or human-induced, occurring per generation is called the *mutation rate* or *mutation frequency.* In evaluating the genetic effects of man-made radiation we customarily speak about the increased mutation rate engendered by such radiation, relative to the natural mutation rate. This is called the *relative mutation risk.*

As stated before, radiation exposure gives rise to changes in *genes* (e.g., deletions, changes in DNA molecule) called *mutations,* and *chromosomes* (e.g., breaks, fusions, altered number, rearrangement of genes) called *aberrations.*

As yet we have no experimental basis for relating the incidence of complex inherited disease to gene mutation frequency in humans. However, we still need at least an estimate of the genetic risk associated with ionizing radiation at low doses so as to set up guidelines for permissible exposure levels. Such estimates must be based either on experimental evidence at the lowest possible doses and dose rates, or on extrapolation from data at high doses and dose rates. Insofar as genetic changes are concerned, a purely *linear dose-response curve* is used for extrapolation. Furthermore, there is good evidence that the incidence of gene and chromosome effects is proportional to dose at low dose levels.

In the BEIR Report (1980) two methods have been used to assess the increased incidence of genetic disorders resulting from gene mutations: indirect and direct. These will now be described.

Indirect Method. Based on gonadal exposure in mice, the indirect method of estimating genetic risk provides a relative mutation-risk factor for radiation delivered during each generation *for a number of generations*:

$$\text{relative mutation risk} = \frac{\textit{mutation rate per rem}}{\textit{spontaneous mutation rate}} \qquad (1)$$

According to the BEIR Report, background radiation causes about 1 to 6 percent of all spontaneous mutations in humans. We may, therefore, assume that a very small increase in exposure above background causes only a small increase in mutation rate, that is, a small relative mutation risk.

For the indirect type of estimate, the relative mutation risk has been derived from experiments with mouse spermatogonia and oocytes. Since the radiosensitivity of spermatogonia is more than twice that of oocytes, a weighted average for the two was obtained and then adapted to humans. On this basis, the *relative mutation risk* turned out to be 0.02 to 0.004 (2 to 0.4%) per rem. The reciprocal of this value, *50 to 250 rems*, is then the *doubling dose*, defined as the dose that would eventually increase the natural mutation rate by 100 percent at genetic equilibrium (see below). In other words, the doubling dose ultimately doubles the natural mutation rate, or induces the same number of additional mutations as occur naturally.

With *dominant genes* the incidence of a genetic trait is proportional to the relative number of mutant genes for that trait—an increase in their number will increase the probability of a detectable defect in the individual. Over a span of generations, not only are mutant genes being induced but they are also being eliminated at a rate proportional to their frequency. Thus, an equilibrium (balance) will eventually be reached between the rate of increase and the rate of elimination. At equilibrium, for example, with a relative mutation risk of 0.01 (i.e., 1% above the natural mutation rate) continued over a number of generations, the incidence of disorders maintained by this mutation would tend to reach 1 percent above the original spontaneous mutation rate. With *recessive* mutations, equilibrium requires many more generations than with dominant mutations because of the much longer time involved in eliminating recessive genes.

How do such estimates of gene mutational frequency relate to detectable genetic disorders? The BEIR Committee estimates an *increase* of 60 to 1100 genetic defects per million live births per rem of parental exposure before conception in the preceding generations. The spontaneous incidence of human genetic defects is now about 107,000 per million live births. Thus the relative increase in human genetic disorders lies within the range of 60/107,000 to 1100/107,000 or *0.05 to 1* percent genetic *disorders per rem*.

Direct Method. In this approach, too, mouse data have been adapted to humans. The direct method of estimating genetic risk tracks the frequency of induced genetic disorders in the first generation after exposure of the parents. Based on the incidence of radiation-induced gene mutations in mouse spermatogonia, calculations were made for all organs and adjusted for human populations. The type of reasoning used in this process may be found in the BEIR report (1980) and is summarized here:

   a. For protracted exposure of mouse *spermatogonia*, the frequency of subsequent skeletal mutations is about 4 per rem per million spermatogonia.
   b. Skeletal mutation frequency is estimated at about 10 percent of all other organs combined.
   c. Therefore, *total* mutation frequency is about 10 times the skeletal rate, with a range of 5 to 15 times.
   d. About 50 percent of genetic defects are significant handicaps, with a range of 25 to 75 percent.
   e. From the foregoing estimates, we can summarize the following computations:

   4 excess skeletal mutations/rem/million spermatogonia. Multiplying by 5 to 15 for all other organs,

   20 to 60 mutations/rem/million spermatogonia. Multiplying by 25 to 75 percent for significant handicaps,

   5 to 45 mutations/rem/million spermatogonia. But oögonia are about 44 percent as sensitive as spermatogonia to irradiation, so multiply by 1.44 to average the effects of irradiating both kinds of gametes, but keep the lower limit the same, giving an excess risk of *5 to 65 induced serious dominant genetic disorders per million liveborn children in the first generation after exposure of the entire population to one rem per generation.*

Baker (1980) gives an interesting example of how the risk factor may be applied in practice. Achondroplasia, a skeletal disorder caused by a *dominant* gene mutation, occurs naturally in about one per 10,000 live births. Doubling the mutation rate of the gene for this disorder (theoretically, 50 to 250 rems) for just *one generation* would increase the incidence of the disease to nearly 1 in 4000 (not quite double the natural incidence because of elimination of some of the genes), after which the excess would gradually decrease to the original incidence in about five or six generations. However, if the doubling dose were continued to each succeeding generation, the mutation rate would rise to 1 in 5000, or twice the spontaneous incidence, in about four or five generations when equilibrium will have been established between the inception of additional mutations and the elimination of mutant genes for achondroplasia.

For *recessive* mutations, a doubling dose to *one generation* would increase the frequency of the resulting disorder to less than 1 percent in the next generation, and even this hardly discernible defect would disappear in the next few generations. However, if the doubling dose were received by each succeeding generation, a doubled incidence of the disease would not be reached for as many as 30 to 50 generations, each of which had received the doubling dose.

TABLE 14.06

GENETIC EFFECTS OF AN AVERAGE POPULATION EXPOSURE OF ONE REM PER 30-YEAR GENERATION (*BEIR 1980*).

(Approximate Values)

Type of Serious Genetic Disorder	Spontaneous Incidence per Million Liveborn Offspring	Effect of One Rem per Generation per Million Liveborn Offspring	
		First Generation	*At Equilibrium* *
autosomal dominant† and X-linked‡	10,000	5–65	40–200
irregular inheritance§	90,000		20–900
recessive	1,100	very few	very slow increase
chromosomal ‖ aberrations	6,000	fewer than 10	increase slightly

*After a number of generations, when the rate at which new mutations appear equals the rate at which previous ones disappear.

†Genes in chromosomes other than those in the X-chromosome.

‡Genes in X-chromosomes.

§Probably result from interplay of a number of different genes, with or without environmental effects.

‖Only those congenital malformations resulting from unbalanced fragments of translocations, or from numerical aberrations.

Note that about 1 to 6 percent of genetic defects probably result from background radiation.

In Table 14.06 are shown the genetic effects of *one rem per 30-year generation* (BEIR Report, 1980) both for gene mutations and chromosomal aberrations. Note that with a natural incidence of about 10,000 autosomal dominant genetic abnormalities per million liveborn children (i.e., 1%). An average population exposure of one rem in the preceding generation (30 years) would yield an excess incidence of 5 to 65 genetic disorders per million liveborn offspring in the first generation. The equilibrium excess incidence is 40 to 200 cases. This value was obtained by multiplying 10,000 by the relative mutation-risk factor 0.004 to 0.02 (0.4 to 2%) per rem. Also note that chromosomal aberrations account for fewer than 10 per million liveborn offspring, a relatively small incidence compared to that for gene mutations.

What would be the effect on the relative mutation rate from a dose equivalent of *5 rems per generation* (30 years), the presently accepted limit for large populations? The answer is obtained by multiplying the dose equivalent by the relative mutation-risk factors:

$$5 \ rems \times 0.004 \ per \ rem = 0.02 \ or \ 2\%$$
$$5 \ rems \times 0.02 \ per \ rem = 0.10 \ or \ 10\%$$

Thus, 5 rems per generation would be expected eventually to increase the mutation rate by 2 to 10 percent.

Unfortunately, we cannot estimate with any degree of certainty the impact of an added 10 percent to the natural mutational burden in humans. We do know that natural background radiation in the United States averages about 100 mrems (0.1 rem) per year, unshielded, and that this has not had any obvious deleterious effect on life on our planet. Furthermore, the natural background level varies from place to place. For example, it is 15 to 35 mrems along the Gulf and Atlantic Coasts (sea level), as high as 140 in Colorado, and even higher in places such as Kerala, India, where it is 2000 mrems per year. Yet, there is no perceptible difference in the incidence of genetic abnormalities.

It is of even greater interest that offspring of 71,000 pregnant women surviving the atomic bombing in Japan have thus far shown no evidence of an effect on six selected genetic factors, but since radiation-induced mutations are almost always recessive, it may take several generations before a firm conclusion may be reached. In any case, there has been no detectable rise in incidence of anomalous children of parents exposed to atomic bomb radiation at an estimated dose of about 100 rems (100,000 mrems). Note that this must necessarily ignore the unanswerable question as to the incidence of embryonic death in early pregnancy.

Nevertheless, every effort should be made to keep gonadal exposure during x-ray examinations at the lowest level consistent with medical needs. Various devices should be available in the radiology department, such as gonadal shields, tight collimation, and optimum screen-film combinations and exposure factors.

Chapter XV

POPULATION EXPOSURE
TO IONIZING RADIATION:
HEALTH PHYSICS

Human exposure to ionizing radiation originates from three sources: natural background, internal, and man-made. We have no practicable means of limiting the first two, but man-made radiation lends itself to at least some degree of control. We shall explore each of these sources of human exposure, but first we must become familiar with the units of dosage applicable to population exposure studies.

Units of Dosage
in Population Exposure

As we have already explained, the unit of absorbed dose is the rad, defined as 100 ergs per gram of absorbing matter. While it has served well as a unit of absorbed dose for various kinds of radiation, the rad does not take into consideration the fact that equal absorbed doses delivered by different kinds of radiation do not necessarily produce the same degree of biologic effect. For example, a 100-rad dose of fast neutrons produces a much more damaging effect on tissues than 100 rads from x or gamma rays. When we are dealing with mixtures of radiations having different LET values, we need a dosage unit that takes into account the differences in intensities of the resulting biologic effects; simply adding the absorbed doses in rads would give a misleading estimate of radiation hazard.

To express on a common scale, for protection purposes *only*, the dosage received by persons from radiations of various qualities, the International Commission on Radiation Units and Measurements (ICRU) introduced the quantity *dose equivalent (DE)*. By definition, DE is the product of the absorbed dose by the appropriate modifying factor for that particular type of radiation, and for a specific biologic effect. This modifying factor is the *quality factor (QF)*. The unit of dose equivalent, DE, is the *rem* (1 rem = 1000 millirems).

$$DE = absorbed\ dose \times QF$$
$$rems = rads \times QF \qquad (1)$$

By specifying appropriate values of QF in equation (1) we can convert rads to rems for any given type of radiation. Note that QF varies with radiation quality (LET) and particular tissue effect. QF differs from RBE in that the former is a measure of the *maximum* biologic effectiveness of the radiation in question and is used for protection purposes only.

For x rays, gamma rays, and beta particles (fast electrons) QF has been arbitrarily assigned the value of 1. Therefore, for these low-LET radiations with which radiologic technologists usually deal,

$$DE = absorbed\ dose \times 1$$

and 1 rem = 1 rad *in this particular case*.

On the other hand, QF for fast neutrons, insofar as cataractogenesis is concerned, is 10, so

$$DE = absorbed\ dose \times 10$$

and for 1 rad of neutrons,

$$DE = 1 \times 10 = 10\ rems$$

This means that an energy absorption of 1 rad from fast neutrons in the eye lens is ten times more effective than 1 rad of x rays in producing cataract; or, conversely, it takes ten times more rads of x rays than of fast neutrons to induce cataract.

The QF for alpha particles is about 20. Hence, the DE for 1 rad of alpha particles is 20 rems.

Note again that the DE in rems of different kinds of radiation can be added because *DE represents a common basis of dosage*. For example, if an individual were exposed to a mixture of radiations and received 1 rad from x rays, 1 rad from fast neutrons, and 1 rad from alpha particles, the total absorbed dose would be 3 rads. However, this would seriously underestimate the radiation hazard. By converting these absorbed doses to dose equivalents we would obtain:

x rays	*1 rad × 1*	*= 1 rem*
fast neutrons	*1 rad × 10*	*= 10 rems*
alpha particles	*1 rad × 20*	*= 20 rems*
	total	*= 31 rems*

Thus, the dose equivalent, in this case 31 rems, provides a more realistic measure of the radiation hazard when the QF exceeds the value one. However, under ordinary conditions in the radiology department, exposure is usually limited to radiation with a QF of one so that the MPD may be stated in rads, or even roentgens because 1 R is numerically almost equal to 1 rad.

Natural Background Radiation

A sensitive radiation detector such as a Geiger-Muller (G-M) counter registers the presence of environmental radiation even in the absence of ordinary known sources such as x rays and radio-pharmaceuticals. We call this prevailing radiation *natural back-ground radiation*. All living matter has been exposed to it from the very beginning of life on earth about one billion years ago.

Natural background radiation includes not only external sources, but also radioactive materials located within the body itself. Furthermore, the detector also contains radioactive material in minute quantity, contributing to the meter reading.

Let us now survey the sources of natural background radiation which may be classified as *external* and *internal*.

External Sources

These comprise cosmic radiation, terrestrial radiation, and atmospheric radionuclides.

Cosmic Radiation. Cosmic radiation consists of two types, primary and secondary. The *primary cosmic rays,* originating in outer space, are mainly high-energy protons (2.5 billion electron volts) but also include alpha particles, atomic nuclei, and high-energy electrons and photons. Interactions of primary cosmic rays with atomic nuclei in the earth's atmosphere produce *secondary cosmic rays.* These consist of such entities as mesons, electrons, and gamma rays. Most cosmic rays that reach the earth's surface are secondary in type and are extremely penetrating. Cosmic ray intensity increases with altitude, being about double at 1800 meters (1 mile) that at sea level. Furthermore, the earth's magnetism causes greater cosmic ray intensity at the poles than at the equator.

Terrestrial Radiation. Radiation arising in the earth is the second external source of background radiation. It includes naturally occurring radioactive minerals in the earth's crust, although in varying amounts from place to place. As would be expected, the largest contribution to background exists near uranium, thorium, or actinium deposits. Terrestrial background radiation, therefore, depends strongly on geographic location.

Atmospheric radionuclides. These constitute the third external source of natural background radiation. They arise through the interaction of cosmic rays with stable nuclides in the earth's atmosphere and include mainly carbon 14, hydrogen 3, and krypton 81.

Because of the effect of altitude and geographic location, the average intensity of natural background radiation varies from place to place. Thus, in the United States, the dose equivalent ranges from 15 to 35 millirems (mrems) per year along the Atlantic and Gulf coasts, to 75 to 140 mrems per year on the Colorado plateau. It is even greater in Kerala, India, where it averages about 2000 mrems per year. In general, the average dose equivalent received in the United States from natural background radiation is about 80 mrems per year (corrected for the shielding effect of housing and the body itself), about equally divided among cosmic, terrestrial, and internal sources.

Internal Sources. Radiation sources within the body include naturally radioactive nuclides as well as artificial radionuclides incorporated in the tissues by ingestion and aspiration. The main radionuclides found in the body are isotopes of potassium (^{40}K), carbon (^{14}C), and hydrogen (^{3}H). Residual radionuclides in the body from atmospheric weapons testing include mainly strontium (^{90}Sr) and cesium (^{137}Cs). Finally, we should mention radium, radon, and polonium; while some of these emit alpha particles, their overall radiation output, including also beta and gamma rays, is very low. The dose equivalent from internally deposited radionuclides averages about 25 mrems per year.

Artificial Sources

Included in the artificial sources of ionizing radiation are those resulting from human technical and industrial enterprises: (1) medical and dental, (2) industrial, (3) air travel, (4) transport of radioactive materials, and (5) nuclear weapons testing. In contrast to natural background, we have some measure of control over artificial sources of ionizing radiation, but there must be a continuous balancing of benefit versus risk so as to achieve maximum benefit with the least possible risk.

We shall now discuss the artificial sources of ionizing radiation affecting the general population (BEIR, 1980).

Medical and Dental X Rays

A general name has been applied to this source—*x rays in the healing arts*. These constitute the largest source of exposure to ionizing radiation generated by man-made equipment. Over 300,000 x-ray machines are devoted to the healing arts in the United States. Of these, just over one-half are used by dentists and the remainder by physicians (medical, osteopathic, chiropractic, and podiatric).

Patient Exposure. In 1970, 65 percent of the people in the United States (i.e., about 130 million) had had one or more x-ray examinations. The total number of procedures is truly astounding:

Medical and	
Dental Radiography	*59 million*
Fluoroscopy	*9 million*
X-ray Therapy	*0.4 million*

In 1964 the Bureau of Radiological Health (BRH) emphasized the *genetic* effects of radiation in terms of the *genetically significant dose (GSD)* as an index of potential harm (see page 191 for the GSD concept). However, with the realization that low-dose radiation can also cause *somatic damage*, the BRH in 1970 decided to use other, additional dose models as biologic indicators; for example, some x-ray examinations may contribute very little to the GSD while exposing the bone marrow to significant doses.

In 1970 the BRH estimated the GSD to be about 20 mrems per year. (Recall that the GSD is the average dose equivalent that, when received by the entire population, would be expected to produce the same total genetic effects as the doses actually received by the exposed individuals; that is, it is the dose received by exposed persons, averaged over the entire population, and adjusted for the expected number of offspring.)

Insofar as *bone marrow dose* is concerned, the United States population received an average of 103 mrads in 1970. Of this 77 percent came from medical radiography, 20 percent from fluoroscopy, and 3 percent from dental examinations.

Occupational Exposure in the Healing Arts. Based on film badge data, the estimated dose equivalent to approximately 195,000 persons operating medical x-ray equipment (i.e., radiologic technologists) averaged 320 mrems in 1968. For the 171,000 persons operating dental x-ray machines, the average dose equivalent was 125 mrems in 1968, decreasing to 50 mrems in 1975. (Note that the annual occupational MPD is 5000 mrems.)

Radiopharmaceuticals. Approximately 10 to 12 million doses of radiopharmaceuticals are administered annually in the United States for medical diagnosis. As would be expected, about 80 percent is represented by agents labeled with technetium 99m; the main organs examined include brain, liver, bone, lung, thyroid, kidney, and heart, in decreasing order. Furthermore, a pilot study by the BRH (1975) shows an annual growth rate of 16 percent in the use of radiopharmaceuticals. The EPA Office of Radiation Programs has estimated that the whole body dose to patients from radionuclides represents about 20 percent of the total resulting from medical diagnostic radiology.

Film badge data for 1975 (National Council on Radiation Protection, quoted by BRH) indicate an annual average dose equivalent of 350 mrems for hospital x-ray and nuclear medicine personnel.

Industrial Sources

We are concerned here with about 70 nuclear power reactors (for generation of electricity), 73 nonpower reactors in research, 80 nonpower reactors operated by the U.S. Department of Energy, and 174 reactors in operation or under construction by U.S. military services (mainly nuclear submarines and surface ships). Thus, we have at present about 400 nuclear reactors of one type or another, either in operation or under construction. Naturally, these involve many steps, starting with mining and purification of uranium, preparation of reactor "fuels," operation of the reactors, and disposal of high level radioactive wastes. One would anticipate occasional malfunction or accidents that could jeopardize personnel as well as populations at various distances from the reactors. Finally, there is the problem of the ultimate disposal of radioactive wastes that are now being stored under questionable conditions of radiation safety.

The Nuclear Regulatory Commission has set out regulations ("regs") limiting the whole body dose equivalent from radionuclides to 8 mrems per year for the general population, and 15 mrems per year to certain organs such as the thyroid gland. For planned release from *all* nuclear reactors in the industry, the whole body limit is 25 mrems per year, and the thyroid 75 mrems per year for the general population. At present, the estimated total environmental release from all nuclear operations averages well below these limits and actually amounts to less than 1 mrem per year.

Airline Passengers and Crew

An often overlooked source of population exposure is the increased intensity of cosmic radiation at the altitudes prevalent in modern air travel. In 1973 about 280 million people in the United States traveled on domestic flights. About 25 percent of the adult population, that is, 35 million individuals, flew at least once in 1973, but each passenger averaged about 10 flights during that year. The average altitude was about 9.5 km (5.5 mi). On the average, each passenger received 3.8 mrems, and the crew 158 mrems in 1973.

Transport of Radioactive Materials

In 1977 there were 2.5 million shipments of all radionuclides, medical and industrial; this represents a significant actual or potential source of population exposure. The U.S. Regulatory Commission estimates that in 1975, 7 million air passengers received an average dose equivalent of 0.34 mrems, 40,000 cabin attendants 3 mrems, and crew 0.53 mrem. Ground crew and by-standers incurred a dose equivalent of 85 mrems as a maximum.

Nuclear Weapons Testing

In the 1950s and 1960s nuclear weapons testing in the atmosphere yielded large quantities of radionuclides that became distributed throughout the world and gradually dropped to the earth as *fallout.* Following adoption of the Limited Nuclear Test Ban Treaty (1963) such testing was severely curtailed, but small amounts of radioactive fallout still persist, acting as a source of population exposure. Furthermore, as we have mentioned above, minute residual deposits of certain radionuclides, especially strontium 90 and cesium 137, remain in the skeletal systems of the population at large, having been absorbed during the fallout process. Certain countries such as the Peoples Republic of China are still conducting atmospheric nuclear weapons testing, thereby adding to the environmental and consequent population burden of radionuclides. As estimated by the U.S. Office of Radiation Programs, the United States population exposure to radioactive fallout to year 2000 will average about 4 to 5 mrems per year.

Table 15.01 summarizes the important sources of U.S. population exposure as recorded by the BEIR in its 1980 report.

PERMISSIBLE EXPOSURE LEVELS

Maximum Permissible Dose Equivalent (MPD)

In view of the many sources of ionizing radiation to which we are all exposed, it becomes obvious that upper limits must be set for population exposure to man-made radiation. (Recall that we

TABLE 15.01

ANNUAL DOSE RATES FROM IMPORTANT SIGNIFICANT SOURCES OF RADIATION EXPOSURE IN THE U.S. (*BEIR 1980*).

Source	Exposed Group	Body Portion Exposed	Average Dose (exposed persons)	Pro-rated over Population
			mrems/yr	mrems/yr
NATURAL BACKGROUND				
cosmic radiation	total population	whole body	28	28
terrestrial radiation	total population	whole body	26	26
internal	total	gonads	28	28
sources	population	bone marrow	24	24
		Total (rounded) 80		80
MEDICAL X RAYS				
Patients				
medical diagnosis	adults	bone marrow	100	77
dental diagnosis	adults	bone marrow	3	1.4
		Total 103		78.4
Personnel (occupational)				
medical	adults	whole body	320	0.3
dental	adults	whole body	50	0.05
RADIOPHARMACEUTICALS				
*Patients (diagnosis)**	adults	whole body	20	13.6
Personnel	adults	whole body	260	0.1
ATMOSPHERIC WEAPONS TESTING	total population	whole body	4–5	4–5

*The Environmental Protection Agency (Report ORP/CDS 72–1) estimates the exposure of patients during nuclear medicine procedures to be about 20 percent of the total radiation exposure in medical diagnosis.

cannot control natural background radiation.) This standard, *the maximum permissible dose equivalent (MPD), may be defined as the maximum dose equivalent of ionizing radiation which an individual may accumulate over a long period of time, or in a single exposure, and which carries a negligible risk of causing significant somatic or genetic damage.* It must be emphasized that the MPD implies the weighing of benefits against risks of radiation exposure; it does not specify "no risk" but rather *"negligible risk."*

The actual numerical value of MPD depends on the segment of the population in question. Thus, there are three *MPD* values applicable to the following situations: (1) occupationally exposed persons, (2) limited portions of the general population, and (3) a large part of the population.

Occupationally Exposed Persons. As applied to technologists, radiologists, and other radiation workers. the lifetime accumulated MPD to the whole body from all sources of ionizing radiation shall not exceed the following:

$$MPD = 5(N - 18) \text{ rems} \tag{2}$$

where N is the attained age and is greater than 18 years. For example, at age 30 years the MPD = $5(30-18)$ = 60 rems. Note that equation (2) reduces to an *annual MPD of 5 rems.* An accumulated dose equivalent of 3 rems is permissible in any 13-week period (i.e., one-fourth year), but only 2 additional rems are then permissible during the remainder of the year. Figure 15.01 shows the occupational MPDs for specific anatomic regions.

In medical radiology, personnel are ordinarily at risk for exposure to x and gamma rays with QF = 1, so the rad and rem are virtually equal numerically in this special situation. Therefore we can express radiation exposure levels in roentgens for radiologic personnel.

Limited Portions of the General Population. For small population segments such as those occasionally receiving radiation exposure, and for pregnant women, the MPD is one-tenth the above value, or *0.5 rem per year.*

Large Population. For the population as a whole, or for large segments of the population, the MPD is limited to *5 rems per generation (30 years),* or *0.17 rem per year.*

The different values of MPD are summarized in Table 15.02. It cannot be too strongly emphasized that the quoted MPDs are

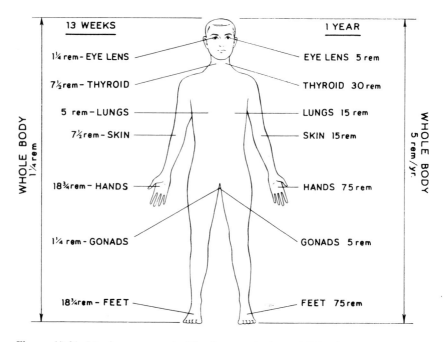

Figure 15.01. Maximum permissible dose equivalent (MPD) for various organs and the whole body of radiation workers, for 13 weeks (left column) and for one year (right column).

maximum values and every effort should be made to limit exposure as far below the MPD as is achievable.

Protection of the Patient in Radiography

As we have already pointed out several times, all x ray exposures, regardless of how small they may be, entail some degree of risk to the patient or his/her offspring. In other words, there is no known threshold dose for radiation injury. In the low-dose region existing in medical radiography, the major risks are stochastic and include carcinogenesis, injury to embryo or fetus, and genetic damage.

Still, this should not deter us from using x rays when medically indicated; but we must be aware of the benefit/risk ratio in each

TABLE 15.02

VALUES OF MAXIMUM PERMISSIBLE DOSE EQUIVALENT (MPD)
FOR THE THREE MAIN POPULATION GROUPS.

Group	MPD
	rems/yr
occupational	5
limited population	0.5
pregnant women	0.5*
general population	0.17†

*MPD for entire period of pregnancy
†Based on 5 rems per 30-year generation.

instance, especially in the fertile population and in infants and children. All necessary steps must be taken to minimize exposure, particularly to the gonads and pregnant uterus, during a radiologic examination. If possible, the number of exposures should also be reduced under these conditions.

Owing to the urgency of protecting the general public from unnecessary radiation, the U.S. Congress in 1968 passed the Radiation Control for Health and Safety Act, which set up the Bureau of Radiological Health (BRH) under the Food and Drug Administration (FDA). The BRH has the responsibility, through its five divisions, of (1) establishing manufacturing standards for all x-ray producing electronic equipment, (2) following up on compliance with these standards, (3) studying the biologic effects of radiation, (4) developing programs for training and medical applications, and (5) overseeing the field of radioactive materials and nuclear medicine.

In 1964, and again in 1970, the BRH surveyed a number of installations and interviewed many individuals to ascertain the trends in patient exposure during x-ray diagnostic procedures. The results appear in summary form in Table 15.03. Note the 20 percent increase in utilization of diagnostic x rays during the 6-year interval. Also, while the average ratio of beam area to film area had dropped 30 percent, the average skin exposure for a plain radiograph of the abdomen had increased by 25 percent.

TABLE 15.03

SELECTED RESULTS OF THE X–RAY EXPOSURE STUDIES
OF THE BUREAU OF RADIOLOGICAL HEALTH.

	1964	*1970*	*Change*
# having 1 or more examinations	108×10^6	130×10^6	+20%
# radiographs per examination	2.2	2.4	+10%
mean ratio of beam area to film area	1.9	1.2	−30%
av. skin exposure per film (AP or PA abdomen)	480 mR	620 mR	+25%
av. skin exposure per film (PA chest)	28 mR	27 mR	0
genetically significant dose per year	16 mrads	20 mrads	+25%

Average entrance exposures for several typical x-ray examinations have been extracted from the 1970 Survey of the BRH and included in Table 15.04. More important are the *gonadal doses* shown in Table 15.05; these were deliberately selected from the 1976 Report of the Food and Drug Administration, Department of Health, Education, and Welfare. The data were obtained retrospectively by survey and do not specify the type of film-screen imaging system. Each radiographic unit should be individually calibrated for R output at the skin level and the dose determined in critical organs based on the imaging system in use (see Rosenstein, 1976). All radiologic technologists must know and constantly keep in mind the dosage in critical organs, such as fertile gonads, when they are included in the radiation field.

The genetically significant dose (GSD) has actually shown an increase from 16 mrads to 20 mrads, or 25 percent in the six-year interval between 1964 and 1970 (see Table 15.03). While this may not be statistically significant, neither does it show a decrease in GSD despite efforts to expand the use of gonadal protection in radiography. Because genetic damage may attend even very small doses of radiation (recall nonthreshold response), gonadal expo-

TABLE 15.04

ESTIMATED AVERAGE EXPOSURE PER FILM AT SKIN ENTRANCE,
BASED ON ALL TYPES OF FACILITIES.*

Examination	Entrance Exposure per Film
	mR
Skull	330
Chest	44
Chest (PA view)	27
Shoulder	260
Thoracic and Cervical Spine	980
Abdomen — KUB	670
Cholecystography and Cholangiography	620
Lumbar Spine	1,920
Lumbosacral Spine	2,180
Upper G.I. Series	710
Barium Enema	1,320
Pelvis	610
Hip	560
Femur	120
Tibia and Fibula	40
Ankle	140
Foot, Toes, Heel	120
Forearm and Elbow	60
Hand and Wrist	100

*From *Population Exposure to X Rays, U.S. 1970.* Bureau of Radiological Health, U.S. Department of Health, Education, and Welfare. Unless otherwise stated, these are mean exposures for various views.

sure must be minimized by appropriate collimation, filtration, and gonadal shielding.

Under the Radiation Control for Health and Safety Act, standards for medical x-ray equipment went into effect on August 1, 1974. Of major importance is the recommendation that the components of a diagnostic unit must, together, make up an integrated system. Furthermore, the manufacturer must test and adjust the equipment at the factory as the final step in the production line. Finally, the installation of the equipment must follow closely the manufacturer's instructions. Some of the more important specifications follow:

TABLE 15.05

ESTIMATED AVERAGE GONADAL DOSE PER EXAMINATION
FROM RADIOGRAPHY BY TYPE OF EXAMINATION AND
BY SEX IN THE UNITED STATES
IN 1970 SURVEY (REVISED).*

Type of Examination	Male	Female
	mrads	mrads
Skull	—	—
Cervical Spine	—	—
Chest	—	1
Thoracic Spine	3	10
Upper G.I. Series	1	170
Barium Enema	175	900
Cholecystography	—	80
Excretory Urography	200	600
Abdomen (KUB)	100	220
Lumbar Spine	220	720
Pelvis	360	210
Hip	600	125
Upper Limbs	—	—
Lower Limbs	15	—

*Data rounded off, based on X-ray Exposure Study (XES), new dose model, Bureau of Radiological Health. The published data include a standard error for each average, amounting to about ±8 to 15 percent.

1. Ability of the equipment to reproduce an exposure for any permissible combination of kV, mA, and time.
2. Proportionality between exposure and time.
3. Provision for mandatory automatic collimation for each film size in stationary x-ray machines.
4. Automatic collimation of the fluoroscopic beam at the patient's entrance surface for spotfilm radiography.
5. Limitation of the fluoroscopic beam within the area of the fluoroscopic screen.
6. Built-in filtration to remove radiation that is too low in energy to be radiographically useful.

Although the actual population exposure seems to be reasonably small, every effort should be directed to reducing it to as low a level as possible. This is especially important for certain radiosensitive structures such as red marrow (blood-forming) and gonads. Protection of the gonads can be achieved by lead shielding and by tight collimation whenever this does not interfere with the examination. Women who may be pregnant should receive special attention: (a) is the examination really necessary? If so, (b) shield the pelvis insofar as possible, (c) limit the number of exposures to the minimum needed for diagnosis, (d) use a fast screen-film system.

At the same time, collimation should serve to minimize somatic injury, especially to the red marrow. Table 15.04 gives representative exposures for various types of examinations.

Exposure can be reduced significantly by adequate beam filtration; thus, above 70 kV a total (inherent plus added) filter of 3 mm Al equivalent reduces skin exposure to about one-fourth that without a filter (Trout *et al.*, 1952). Patient exposure can be further decreased by an increase in kV with appropriate decrease in mAs to maintain constant radiographic density.

The important factors in reducing radiographic exposure of the patient—*minimum dose radiography*—may be summarized as follows:

1. Filtration of beam by a total of 3 mm Al equivalent.
2. Tight collimation of beam.
3. Lead shielding of gonads whenever possible.
4. Film-screen combinations with highest speed consistent with good radiographic quality.
5. Optimum automatic film processing.
6. Optimum kilovoltage technic.
7. Careful positioning and choice of technic to avoid unnecessary re-examinations.

In *fluoroscopy*, the recommended factors for achieving minimal patient dose include the following:

1. Filtration of beam by a total of 3 mm Al equivalent.
2. Optimum technical factors: 85 to 100 kV, 2 to 3 mA; *kV reduction for infants and children.*
3. Collimation to smallest area appropriate for the examination.
4. Focus-to-tabletop distance no less than 30 cm (12 in.) and preferably 40 cm (15 in.).
5. Exposure rate at tabletop not to exceed 10 R per min.

In summary, we should again emphasize that the total exposure for medical diagnosis should be kept at the lowest level, both somatic and gonadal, consistent with medical needs. A permanent record should be kept for each exposure, including the region exposed, kV, mAs, and filtration. This serves not only to keep track of radiation exposure, but also to facilitate duplication of radiographic exposure on follow-up examinations. Again, bear in mind that strict limits must be placed on radiography of pregnant women as to number of exposures, collimation, and lead shielding to protect the radiosensitive fetus. As noted above, the total dose to the fetus must be kept as far as possible below 0.5 rad during the entire period of gestation, except where the benefit to the mother far exceeds the risk to the fetus.

Protection of Personnel
in Radiography

Let us now turn to the procedures that enable us to keep personnel exposure well below the occupational MPD of 5 rems per year. We shall consider these under four headings: radiation monitoring, wall protective barriers, working conditions, and protection surveys.

Radiation Monitoring

We can monitor whole-body exposure without special equipment or training. (A qualified health physicist can do this more precisely, and should at least perform radiation safety surveys at prescribed intervals.) Several devices are available, including (1) pocket dosimeter, (2) film badge, and (3) thermoluminescent dosimeter (TLD).

Pocket Dosimeter. Outwardly resembling a fountain pen, the pocket dosimeter contains a thimble ionization chamber at one end. The exposure in R can be read by means of an auxiliary electrometer, or, in the self-reading type, directly by means of a built-in electrometer. Such dosimeters, while sensitive to exposures up to 200 mR, are unreliable in inexperienced hands, do not furnish a permanent record, and are too fragile for general use.

Film Badge. This is at present the most convenient and practicable type of personnel monitor. Commercial laboratories specialize in supplying and servicing the film badges, and maintaining permanent records. The x-ray film badge consists of a small piece of special film covered by various filters to show the quality of the radiation and to permit conversion to tissue dose. The film is also backed by a thin sheet of lead to absorb scattered radiation from behind the badge. The entire assembly is enclosed in a light-tight packet and worn on the front of the body. During fluoroscopy, it should be worn outside the apron at the front of the neck to monitor exposure of the thyroid gland and eyes.

The laboratory supplies a fresh badge monthly for each member of the radiologic staff, including all persons who may be regularly or frequently exposed to radiation. After being worn, at first for one week, the badge is returned to the laboratory where the film is processed under standard conditions and compared densitometrically to standard films that have previously received a known radiation exposure. The dose equivalent in rems is reported to the radiology department for review and permanent filing.

Badges should never be exchanged among personnel and should be clearly marked for identification. With exposure levels well below the weekly MPD (0.1 rem) the film badge should continue to be worn and returned at monthly intervals especially if the work load is reasonably constant. The one-month period is preferred because of convenience, greater accuracy of calibration, and lower cost.

Thermoluminescent Dosimeter. Relatively new on the scene, the TLD type personnel monitor works on the principle that certain crystals such as lithium fluoride (LiF) can store energy on exposure to ionizing radiation, which causes valence electrons to be trapped in crystal lattice defects. When the crystals are heated under strictly controlled conditions, the trapped electrons return to their normal state accompanied by a release of energy in the form of light. Measurement of the emitted light can then be related to the initial radiation exposure. The dosimeters are usually returned to a commercial laboratory for readout. Lithium fluoride (LiF) is the most commonly used detector material.

TLD detectors have a number of advantages over film badge monitors:

1. Relatively inexpensive.
2. Wide exposure range detectable—1 mR to 1000 R, response being proportional to exposure up to about 400 R.
3. Response to photon radiation nearly independent of quality in the range of 50 kV to 20 MV.
4. Response similar to that of tissue, that is, TLD is virtually tissue equivalent.
5. Accuracy ± 5% (film badge ± 50%).
6. Available in very small format, for example, 1 mm × 6 mm.
7. Feasibility of incorporation in jewelry such as finger rings for nuclear medicine technologists, and pins that can be worn on the clothing.

For these reasons, it is very likely that TLD monitors will eventually replace film badges for personnel monitoring.

On termination of employment all technologists must receive an appropriate summary of occupational exposure regardless of the monitoring system, provided it has been approved by the licensing agency. If reemployed, the technologist should present the summary to the new employer.

Wall Protective Barriers

This requires built-in protective barriers of suitable radiation-attenuating material to reduce radiation exposure below the MPD. Wall protective barriers must be planned in advance by a *qualified radiation physicist*.

An important principle in wall barrier design is that wherever the wall may be struck by the useful beam (i.e., the direct radiation from the source) a greater thickness of a particular material must be used than where it is struck by scattered or leakage radiation (leakage refers to radiation that has penetrated the housing around the tube or source, other than the port). Furthermore, all joints and holes must be covered by the same or equivalent protective barrier as the wall itself. Adequate wall protection varies with beam energy as described in National Council on Radiation Protection and Measurements (NCRP) Handbook #33.

Terms that often apply to wall protection are "controlled" and "uncontrolled" areas. A controlled area is one that is under supervision of a Radiation Safety Officer, whereas an uncontrolled area is not.

Working Conditions

Despite the adequate wall protection that prevails in modern radiology departments, the personnel must follow certain rules if they are to keep their exposure well below the MPD. These should be impressed repeatedly on all radiology department personnel, since their cooperation is essential for their own safety.

1. Never expose a human for demonstration purposes alone.
2. Never remain in a radiotherapy room while a treatment is in progress; approximately 0.1 percent of the useful beam is scattered perpendicular to the beam at a distance of 1 meter.
3. Only occasionally should you hold a patient in radiography, since many efficient holding devices are available both for patient and cassette. If a patient must be held for radiography, this should be done whenever possible by a person not habitually exposed to ionizing radiation; in any case, a lead apron and gloves are mandatory.
4. Never hold a cassette for horizontal cross-table radiography; a lead apron offers minimal protection against a direct beam. Use a cassette-holding device.
5. In fluoroscopy, wear a protective apron usually made of lead-impregnated vinyl with 0.5 mm lead equivalent and, when not actually assisting the fluoroscopist, stand either in the control booth or behind the fluoroscopist.
6. Check lead-protective gloves and aprons periodically for cracks, using a radiographic test with par speed screens and these exposure factors: 100 kV and 10 mAs at 100 cm (40 in.) focus-film distance. For other types of screens adjust the *mAs* in accordance with screen speed.

Protection Surveys

The evaluation of actual or potential radiation hazards in a radiology department requires a *complete protection survey* and *periodic resurveys* by a qualified expert. Should new equipment be added, old equipment modified, or procedures changed, a resurvey becomes mandatory. More barrier protection should then be added as required, followed by a resurvey after such modification.

If, in the opinion of the qualified expert, there is reasonable probability that a person in a controlled or uncontrolled area could receive more than *10 mR in one week*, then one or more of the following steps must be taken:

1. Determine the cumulative dose in the area in question.
2. Use personnel monitoring in the area in question.
3. Add barrier material to equipment or walls to comply with authoritative recommendations in NCRP Report #34.
4. Impose restrictions on the use of the equipment, or on the direction of the beam.
5. Impose restrictions on the occupancy of the area if this is controlled.

All "on-off" control mechanisms (control panel, entrance door, emergency cutoff switch) must be inspected *semiannually* and repaired if necessary.

Warning signs must be posted in a prominent place in the appropriate areas as follows:

1. CAUTION RADIATION AREA sign in area where prevailing exposure rate is more that 5 mR per hour, but less than 100 mR per hr.
2. CAUTION HIGH RADIATION AREA sign at entrance to any area in which the exposure rate is 100 mR per hr.
3. CAUTION RADIOACTIVE MATERIAL sign where the activity exceeds a specified level — ^{131}I, 10 µCi; ^{32}P, 100 µCi; ^{60}Co, 10 µCi.

Whenever any restriction has been placed on equipment, such as limitation of beam direction to prevent penetration of an inadequate barrier, the specified restriction must be enforced.

All reports of calibrations and surveys should be made in writing, signed by the qualified expert, and filed *permanently*. The report should state if and when a resurvey is needed.

Electrical power to therapy equipment must be locked in the "off" position when not in use. Instructions must be posted conspicuously in the control area describing the steps to be taken in the event equipment cannot be turned off.

The name, address, and phone number of the radiation safety

officer and the state agency concerned with radiation safety must be posted in the control area for prompt service in the event of an emergency. A responsible substitute should be available, on call, if the radiation safety officer cannot be reached.

Protection of the Patient in Nuclear Medicine

The risk to patients undergoing nuclear medicine examinations resembles that from diagnostic x rays at similar dosage levels. However, there is one major difference in that x rays *uniformly* irradiate the radiation field, whereas radiopharmaceuticals concentrate in certain organs called *target organs*. For example, radioactive iodine accumulates selectively in the thyroid gland so that ingestion of 0.05 millicurie of iodine 131 will deliver approximately 100 rads to the thyroid gland, but only 0.2 rad to the whole body. Incidentally, in any radionuclide procedure the body as a whole as well as organs other than the target organ become irradiated. However, in the majority of *in vivo* nuclear medicine procedures the target organ receives a dose about equal to or greater than the rest of the body.

In disease, a nontarget organ (and even the target organ itself) may enlarge or display an increase in function, which may increase the concentration of the radiopharmaceutical above the usual values. Yet, this is seldom a problem in clinical practice.

No specifically increased danger from diagnostic radionuclides has been found relative to that from x rays. Rad for rad, one should anticipate low-dose hazards from radiopharmaceuticals to be the same as those from x rays with similar distribution. However, in actual practice the relatively small radionuclide doses used in diagnostic studies deliver smaller absorbed doses than those prevailing with x ray examinations. Another factor that diminishes the biologic effect of radionuclides is the low dose rate associated with their decay, in contrast to the higher dose rate from x rays, but its practical importance is questionable.

To keep patient dosage as low as possible, much progress has been made toward the development of radionuclides that emit pure gamma rays; beta particles have no diagnostic value since they are almost completely absorbed within the body, but they still contribute to the absorbed dose. Outstanding has been the production of various radiopharmaceuticals containing ^{99m}Tc, a radionuclide that has assumed major importance in diagnosis by virtue of its 140-keV gamma rays. These, together with the accompanying 84-keV beta particles, deliver a relatively small absorbed dose.

Another way to minimize absorbed dose is to use radionuclides having a very short half-life; in this respect, too, ^{99m}Tc with a half-life of six hours serves as an excellent imaging radionuclide. Unfortunately, some other short-lived radionuclides such as iodine 123 (half life 13.2 hours) have extremely limited application because they cannot be milked from a "cow" and therefore have to be used close to the source of supply.

Finally, continual improvement in detector efficiency and data display systems has made possible a decrease in the required radionuclide activity, thereby reducing the absorbed dose in the patient.

Protection of Personnel in Nuclear Medicine

The medical use of radionuclides has undergone phenomenal growth since 1946 when suitable radiopharmaceuticals became commercially available. In fact, nuclear medicine ranks high in the field of medical diagnosis, especially with the constant improvement in equipment and the introduction of excellent short-lived radiopharmaceuticals.

Since about 10 to 12 million doses (BRH Report, 1980) are now being administered annually in the United States, precautions must be taken to avoid unnecessary exposure of patients and personnel. The problems associated with radionuclides differ in many respects from those prevailing in radiography; hence, the need to discuss protection in nuclear medicine under a separate heading.

Types of Radiation

Let us review briefly the types of radiation prevailing in nuclear medicine. Radionuclides give off one or more of the following radiations: alpha particles, beta particles, gamma rays. Figure 15.02 shows the respective penetrating abilities as expressed by half-value layer. Since medical radionuclides emit only gamma rays and/or beta particles, we shall limit our discussion to these two kinds of radiation.

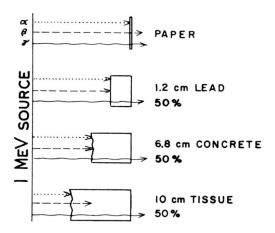

Figure 15.02. Comparative absorption of alpha, beta, and gamma radiation in various materials. Also shown are the half-value layers of 1-MeV gamma rays in lead, concrete, and body tissues.

Not only do gamma rays have a much greater penetrating ability than beta particles, but they also differ from them in character. Gamma rays are photons like x rays, while beta particles are high-speed electrons. Consequently, protection must differ, as we shall see later.

Harmful Effects

These have already been described in detail for ionizing radiation in general. Similar effects are produced by beta and gamma radiation for equal absorbed doses because they are both low-LET radiation. However, beta particles have relatively poor penetrating ability so that with radionuclides used in nuclear medicine effects are limited mainly to the surface with external exposure (e.g., strontium 90), or to the immediate vicinity in the case of internally deposited beta emitters. However, as beta particles slow down in the field of atomic nuclei, brems radiation (x rays) is given off.

On the other hand, the highly penetrating gamma radiation given off by internally deposited gamma emitters can pass out through the surface of the body, and external gamma rays can penetrate deeply into the body.

Maximum Permissible Dose Equivalent

Because the emissions from radiopharmaceuticals include only beta and gamma radiation, the maximum permissible dose equivalent (MPD) can be stated as already explained on pages 207–208. The MPD is the same — 5 rems per year whole body, including the gonads and eyes, of radiation workers. The MPD to the hands and forearms is 75 rems annually. Furthermore, the MPD to the hands in any calendar quarter must not exceed 25 rems, so that only 50 rems may be accumulated the rest of the year (75 rems − 25 rems = 50 rems). This is of particular importance to nuclear medicine (and radium) technologists.

According to the BRH Report (1980), 81 percent of the radiopharmaceuticals used in nuclear medicine are tagged with technetium 99m, which emits 140-keV gamma rays (also very low energy beta particles, av. 84 MeV). The average annual whole body dose equivalent to nuclear medicine technologists in 1968 was 260 mrems; this had increased to an average of 700 mrems (range 150 to 4000 mrems) from ^{99m}Tc generators in 1973 (DHEW Publication, FDA, #73-8029), and may well have increased further since then. In the same publication, the hand exposure of six nuclear medicine technologists from ^{99m}Tc averaged 11 mR per injection, which could result in an *annual* exposure of as much as 15,000 mR (15 R) when unshielded syringes are used. A special lead-shielded syringe reduces the hand dose to about 10 percent of this value, or 1500 (1.5 R) per year.

Principles of Personnel Protection

There are four approaches to reducing or eliminating radionuclide hazards. These include distance, shielding, time of exposure, and limitation of radionuclide activity.

Distance. This relatively simple and inexpensive method of reducing gamma-ray exposure represents an application of the inverse square law. Thus, radionuclides should be stored as far as conveniently possible from occupied areas. Long forceps for picking up sealed sources such as radioactive capsules, needles, and wires, and removing solutions from shipping containers, will significantly reduce the exposure to the hands and forearms. With highly active sources, remote control equipment becomes necessary. In any case, distance helps greatly to reduce personnel exposure below the MPD.

The following equation gives the shortest permissible distance (SPD) for gamma emitters, that is, the distance from a stored or internally deposited radionuclide at which the exposure rate approximates the MPD in a 40-hour week:

$$D = 20\sqrt{A\Gamma}\ cm \tag{3}$$

where D is the SPD, A is radionuclide activity in millicuries (mCi), and Γ (Greek capital gamma) is the specific gamma-ray constant (defined as the exposure rate in R per hr at a distance of 1 cm from a point source of a particular radionuclide having an

TABLE 15.06

SPECIFIC GAMMA–RAY CONSTANTS (Γ) OF SOME RADIONUCLIDES.*

Nuclide	Γ
	R/mCi-hr at 1 cm
^{198}Au	2.3
^{60}Co	13.0
^{131}I	2.18
^{137}Cs	3.3
^{226}Ra (0.5 mm Pt filter)	8.25 (0.5 mm Pt capsule)

*Based on *ICRU Handbook 86* (experimental values).

activity of 1 mCi). For example, what is the SPD from a 10 mCi capsule of iodine 131? The value of Γ for ^{131}I (see Table 15.06) is 2.18 R/hr at 1 cm. Therefore,

$$D = 20\sqrt{10 \times 2.18}\ cm$$
$$= 20\sqrt{21.8}$$
$$D = 20 \times 4.7 = 94\ cm\ (3\ feet)$$

In practice, distance alone is inadequate in a nuclear medicine department so lead shielding must be added.

Shielding. Protective materials placed between a radioactive source and its surroundings are called *shields* or *barriers*. Adequate shielding contributes materially to reducing *gamma-ray exposure* of persons in the vicinity of the radioactive material. Radiopharmaceuticals are shipped in lead-protective containers, and as a rule, should be stored in them. In addition they should be surrounded by lead bricks 5 cm (2 in.) in thickness on all sides that might constitute a radiation hazard. Monitor surveys will then establish barrier adequacy, more lead being added if necessary.

The increase in thickness of any protective material needed for additional radionuclide activity is less than you may think. Suppose we have 100 mCi of a radionuclide for which 4 cm lead provides an adequate barrier. What *additional* lead thickness would be needed for 200 mCi of the same radionuclide? We must first know the half-value layer in lead for this particular radionuclide, which in this case is 1 cm. The addition of only 1 cm lead (making a total of 5 cm) should then provide the same level of protection as obtained initially. You can see this from the definition of half-value layer. Adding 1 HVL will reduce the exposure rate from 200 mCi by 50 percent, or to the same level as that from 100 mCi of the same radionuclide.

Shielding of *beta particles* differs from that of gamma rays. Since the range of beta particles is limited to a cm or so in soft tissue, the external hazard can be readily controlled. The range of beta particles depends on their energy; for example, low energy beta particles (say, about 0.3 MeV) are absorbed by the glass container, the outer layer of skin, or about 100 cm of air. On the other hand, the medical radionuclides phosphorus 32 (max. beta energy 1.7 MeV) and strontium 90-yttrium 90 (max. beta energy 2.18 MeV) require special shielding to protect against brems radiation produced when the high-energy beta particles pass close to atomic nuclei. Plastics such as Lucite® or polystyrene are preferred because they are light-transparent; besides, their lower atomic number (than lead) is associated with less intense brems-ray production. About 6 mm (¼ in.) plastic suffices for ^{32}P, and about 10 mm (⅜ in.) for ^{90}Sr.

Time of Exposure. Since total exposure equals exposure rate times time, you can see that the faster a procedure is carried out, the smaller will be the exposure to the technologist. Each different procedure should be first practiced with blank (nonradioactive) material until sufficient speed has been attained. This applies especially to radiopharmaceuticals of high activity, and to radium and artificial radionuclide sources for brachytherapy.

Limitation of Activity of Stored Radionuclides. Exposure rate is directly proportional to source activity; for example, the exposure rate from 100 mCi iodine 131 is twice that from 50 mCi at the same distance. Obviously, the exposure rates from all sources in the storage area must be added to obtain the total exposure rate. Hence, the quantities stored should be kept at the minimum consistent with the work load.

The following list of unsafe practices serves as a useful indicator of improper work habits in the nuclear medicine department:

1. Inadequate shielding of stored radionuclides and syringes.
2. Inadequate monitoring.
3. Inadequate planning—excessive time spent in conducting procedures.
4. Failure to use trays and to cover work areas with paper when liquid radiopharmaceuticals are used.
5. Pipetting radioactive solutions by mouth instead of remote devices.
6. Poor work habits such as smoking or eating in the laboratory.
7. Failure to wear suitable outer garments such as plastic gloves and laboratory coats when handling high-level radioactive liquids.
8. Improper waste disposal.
9. Improperly designed fume hoods in high-level laboratories.
10. Changing the levels of activity without modifying procedures to maintain low hazard.
11. Failure to keep adequate records of radionuclides received, waste disposal, unusual incidents, and personnel exposure.
12. Failure to post signs indicating level of activity, and storage and disposal areas.

Chapter XVI

RADIATION ONCOLOGY

In its broadest sense, the term *oncology* refers to the scientific study of tumors. Here we shall deal with oncology as it pertains to the diagnosis and treatment of tumors, usually of the malignant or cancerous types. Accordingly, an *oncologist* is a physician who specializes in this field.

Because of the diverse treatment methods for various cancers in different anatomic locations, modern cancer therapy requires a *team approach* involving a number of medical and surgical specialists: radiation oncologist, surgeon (including surgical subspecialists), internist, pathologist, medical oncologist, and hematologist. All should participate to some degree in the planning and execution of a course of treatment designed for a specific situation. Available modalities include radiotherapy, surgery, chemotherapy, hormone therapy, and immunotherapy, and occasional supplementary physical methods such as hyperthermic and hyperbaric therapy.

The *radiation oncologist* (also known as the radiotherapist) has a number of technical associates who jointly help plan and carry out irradiation therapy when indicated. They include mainly the radiation physicist, dosimetrist, and radiotherapy technologist.

Essential to cancer control is the initial determination of the histologic tumor type (by biopsy) and extent of disease (by such steps as inspection, palpation, radiography, radionuclide imaging, computerized tomography, ultrasound scanning, nuclear magnetic resonance, biopsy, and surgery).

Basic Oncology

Before discussing irradiation therapy we should have a clear understanding of oncologic terminology.

The term oncology (*onkos* tumor + *logos* discourse) encompasses

226

all aspects of the nature, diagnosis, and treatment of tumors. Another name for tumor is *newgrowth* or *neoplasm* (*neo* new + *plasm* form), the process of tumor development being called *neoplasia*. Neoplasms comprise two types, *benign* and *malignant*. Although the experienced pathologist can usually distinguish between them by microscopic examination of specially stained thin sections of tumor tissue prepared in a special way (histologic examination), the dividing line between benign and malignant neoplasms is not always well defined. The microscopic features of neoplasms have been further clarified in recent years by means of the electron microscope.

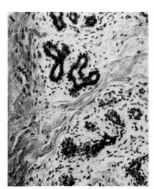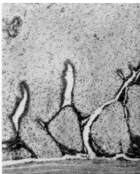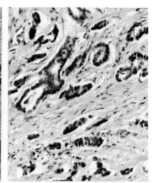

Figure 16.01. Examples of normal and neoplastic breast tissue as seen under the light microscope (in contradistinction to the electron microscope).

A is a section of normal breast tissue. The dark cells arranged in tubules represent glands and ducts in a regular pattern, imbedded in connective tissue stroma. (From W. M. Copenhaver et al [Eds.], *Bailey's Textbook of Histology*, Copyright 1978, The Williams & Wilkins Co., Reproduced by permission.)

B is a section through a *fibroadenoma*, a benign tumor consisting of glandular and stromal tissue arranged in a regular pattern, without mitoses or invasion of adjacent tissue. (From S. L. Robbins, Textbook of Pathology with Clinical Applications, 1962. Courtesy of W. B. Saunders, Philadelphia.)

C is a section through an *adenocarcinoma* (malignant tumor). Note the irregular arrangement of gland-like structures. A higher magnification would show cells in mitosis and invasion of cancer cells into the stroma. (From W. A. D. Anderson, *Pathology*, 1971. Courtesy of C. V. Mosby, St. Louis.)

Benign neoplasms, which characteristically exhibit slowly expanding growth and associated tumor capsule formation, consist of cells having virtually no mitotic activity (see Figure 16.01B).

Such tumors may cause signs and symptoms related to compression of neighboring organs, but do not invade normal tissues. They are best treated by surgical removal, preferably before they reach excessive size. Surgery is also indicated for those benign tumors that tend to become malignant, such as certain cartilage tumors, especially when they become painful or exhibit growth.

Malignant neoplasms (see Figure 16.01C) display unrestrained growth, do not have a tumor capsule, invade (infiltrate) nearby tissues, and metastasize (set up colonies of tumor cells) in distant regions via the blood and lymphatic streams. In addition, they may spread or seed in the pleural or peritoneal cavities. Symptoms and signs of malignant tumors may be specifically related to the organs they invade, or involve by metastasis. Malignant tumors give rise to generalized symptoms such as weakness, fever, anorexia (loss of appetite), weight loss, and general deterioration of health. They ultimately have a fatal outcome if treated unsuccessfully.

Neoplasms may be classified according to two main tissues of origin, *epithelial* and *mesenchymal* (nonepithelial). *Epithelial tumors* arise from epithelium, which includes all the external and internal surface coverings (i.e., linings) of the body, as well as their extensions. For example, the surface linings of the skin, mouth, throat, esophagus, gastrointestinal tract, respiratory tract, genitourinary tract, heart, and blood vessels are all epithelial structures. *Mesenchymal tumors originate* in tissues and organs that stem from embryonic mesenchyme (loose connective tissue), and include tumors of muscle (skeletal, intestinal, cardiac, etc.) and connective tissue.

The term *cancer* applies to *all* malignant tumors. We generally subdivide cancer into two main categories, *carcinoma* and *sarcoma* (see Figure 16.02).

1. *Carcinoma*—any malignant tumor originating in epithelial tissue.
2. *Sarcoma*—any malignant tumor originating in mesenchymal tissue, such as muscle, cartilage, bone and connective tissue.

These groups are further subdivided according to their histologic (tissue) origin. For example, carcinomas resembling skin, although arising from the mouth, esophagus, cervix, or lung, are

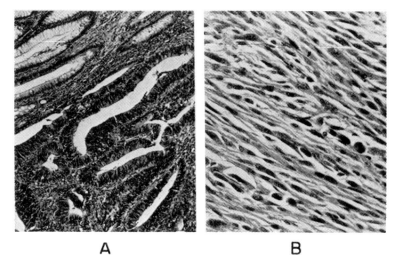

A **B**

Figure 16.02. Examples of carcinoma and sarcoma as seen under the light microscope.

A is a section through an adenocarcinoma of the colon with adjacent normal colon for comparison. In the upper left-hand corner are the normal glands made up of cells with clear cytoplasm and regularly arranged basal nuclei. The remainder of the section shows the adenocarcinoma with its irregular gland-like pattern, closely packed dark-staining cells, and nuclei that vary in size, shape, and position.

B is a section through a fibrosarcoma showing its spindle-shaped cells and their nuclei, which vary widely in size and shape.

(From S. L. Robbins, *Textbook of Pathology with Clinical Applications*, 1962. Courtesy of W. B. Saunders, Philadelphia.)

called *epidermoid* or *squamous cell carcinomas.* Those of gastrointestinal origin most often have a glandlike structure and are therefore called *adenocarcinomas* (see Figure 16.03).

A *sarcoma* of fibrous tissue origin is a fibrosarcoma; muscle, myosarcoma; bone, osteosarcoma; cartilage, chondrosarcoma. Figure 16.04 shows three kinds of sarcomas. These are only samples of the numerous histologic varieties of malignant tumors.

Benign tumors have names ending in *-oma*, without the prefixes carcin- or sarc-. For example, benign epithelial tumors are exemplified by adenomas. Benign mesenchymal tumors include fibromas, myomas, neuromas, osteomas, and chondromas.

Malignant tumors are further classified microscopically as to their state of *differentiation* (maturity). Usually, the pathologist

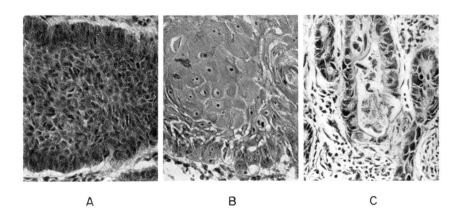

A   B   C

Figure 16.03. Examples of three kinds of carcinomas as seen under the light microscope.

A is a section through a basal cell carcinoma of the skin. Note the irregular size and shape of the cells that line up at the periphery—so-called palisading.

B is a section through an epidermoid (squamous cell) carcinoma that has penetrated into the dermis (see also Figure 8.07).

(A and B from S. L. Robbins, *Textbooks of Pathology with Clinical Application*, 1962. Courtesy of W. B. Saunders, Philadelphia.)

C is a section through an adenocarcinoma of the colon whose glands are made up of cells having irregular shapes and sizes, and dark nuclei many of which are in mitosis. (From H. T. Karsner, *Human Pathology*, 1942.)

indicates whether the tumor is well, moderately, or poorly differentiated and may further grade the tumor from one to four in increasing order of malignancy. While this may have some prognostic value, tumor grade may vary in different parts of the same tumor so that if the pathologist happens to choose at random a less malignant zone of the tumor for histologic examination, he will underestimate its grade.

Clinical Staging

Once the diagnosis has been established by appropriate cytologic (cell) and histologic (tissue) study, the extent of disease must be

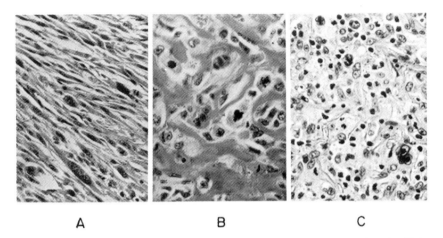

A   B   C

Figure 16.04. Examples of three varieties of sarcomas as seen under the light microscope. These are all malignant tumors.

A is a section through a *fibrosarcoma* with its spindle-shaped cells and its nuclei that vary in size and shape.

B is a section through an *osteogenic* (bone-forming) *sarcoma* showing marked variation in cellular and nuclear size and shape; such tumors characteristically produce a substance known as osteoid, as does normal bone. (A and B from S. L. Robbins, *Textbook of Pathology with Clinical Applications*, 1962. Courtesy of W. B. Saunders, Philadelphia.)

C is a section through a lymph node involved by mixed cellularity *Hodgkin's disease*. Note the nonuniform size and shape of the cells and the occasional, much larger multinucleated Sternberg-Reed cells. Reprinted by permission of the publishers from *Hodgkin's Disease*, by Henry S. Kaplan, Cambridge, Mass.: Harvard University Press, Copyright© 1980 by the President and Fellows of Harvard College.

determined before initiating the treatment plan. Clinical staging, while not an exact science, attempts to accomplish this through careful examination of the patient. The staging becomes a part of the patient's record and is not changed by the subsequent course of the disease. In some instances, supplementary surgical staging may be included. Clinical staging aims to provide as accurate a record as possible of the anatomic site and extent of disease to aid in treatment planning, prognosis, evaluation of treatment results, and intercomparison of results among various treatment centers.

There are two main methods of classification applicable to cancer staging. One that has been in use for many years includes four stages specified by the Roman numerals I, II, III, and IV, in order of increasing size and extent of tumor. Subgroups further refine the staging when appropriate.

The other classification, devised by the *International Union Against Cancer (UICC)*, seeks to improve clinical staging by means of the *TNM system*, wherein T indicates tumor size, N refers to the status of the regional lymph nodes, and M designates the presence or absence of distant metastases. This system applies only to previously untreated cases and must be based only on *clinical examination*, which may include any type of diagnostic radiology and endoscopy. Another TNM system has been developed by the *American Joint Committee (AJC)*, which is not always exactly the same as the UICC system, but steps are being taken to unify them.

As with the other staging system, the TNM system dose not permit changing the stage once it has been established, although it does allow for supplemental classification based on surgical findings.

Table 16.01 gives the TNM staging system for carcinoma of the

TABLE 16.01

INTERNATIONAL UNION AGAINST CANCER (UICC) STAGING OF
CARCINOMA OF THE BREAST. TNM SYSTEM

T0	No demonstrable cancer in the breast
TIS	Preinvasive, in situ carcinoma
T1	Tumor not over 2 cm in greatest diameter
T2	Tumor more than 2 cm but less than 5 cm in greatest diameter
T3	Tumor larger than 5 cm in greatest diameter
T4	Tumor of any size with extension to chest wall or skin
N0	No palpable axillary nodes
N1	Movable homolateral (same side) axillary nodes
	N1a Thought not to contain tumor
	N1b Thought to contain tumor
N2	Fixation of homolateral nodes to each other or to adjacent tissue
N3	Supraclavicular or infraclavicular homolateral nodes or edema (swelling) of the arm
M0	No evidence of distant metastasis
M1	Distant metastasis (e.g., lung, bone, etc) and/or skin beyond breast area

breast. Note the purely clinical basis on which the staging is carried out. In Table 16.02 we have shown how the TNM system may be incorporated into the Roman numeral system.

Let us now see how staging is applied in an actual clinical situation. Suppose we have a patient with a 4 cm breast cancer (T2) with fixed axillary lymph nodes (N2) and no distant metastasis (M0). In the TNM system (AJC) this tumor would obviously be classified as T2N2M0. Referring to Table 16.02 we can see that it would be declared a Stage III lesion because of N2.

TABLE 16.02

CONVERSION OF TNM SYSTEM TO ROMAN NUMERAL SYSTEM OF STAGING CARCINOMA OF THE BREAST.

Stage I—all M0

T1, N0 or N1a (with or without fixation of breast tumor to skin or underlying pectoral muscle, does not change classification)

Stage II—all M0 (with or without fixation of breast tumor to skin or underlying pectoral muscle, does not change classification)

T0N1b

T1N1b

T2N0 or T2N1a or T2N1b

Stage III—all M0 (with or without fixation of breast tumor to skin or underlying pectoral muscle, does not change classification)

T3 with any N (i.e., T3N0 through T3N3)

T4 with any N

Any T with N2

Any T with N3

Stage IV—any stage with M1

An important contribution of the TNM system is its specification of the routes by which a particular tumor has spread. Thus, N indicates whether tumor cells have metastasized via the lymphatics (i.e., to lymph nodes) and M via the bloodstream as well as lymphatics to distant sites. As a general rule, carcinomas are more likely to metastasize to regional nodes first, although they often metastasize widely through the bloodstream as well. They also seed body cavities (pleural and peritoneal) by surface implantation. Sarcomas have a much greater tendency to metastasize through the bloodstream and seldom develop nodal metastasis, although exceptions occur.

Growth of Tumors

Before discussing the use of ionizing radiation in the treatment of malignant tumors we must review the kinetics of cell populations. This subject was introduced earlier (see pages 75–78) with relation to normal tissues and organs and may be extended to malignant tumors. Cell population kinetics deals with the progression of cells from primitive forms to more mature ones, although some tumors consist entirely of immature cells. On the other hand, immature tumor cells may in some cases result from dedifferentiation of mature cells, that is, reversion of such cells to less mature forms.

On the assumption that a tumor starts from one cell and that the net growth rate (difference between cell production and cell loss) is exponential, that is, the fractional increase in volume is constant, we can derive an estimate of tumor growth or tumor duration by using the concept of *observed doubling time*; this is the time it takes for a tumor to double in mass or volume, a concept originated by Collins. In general, a tumor must reach a mass of about one gram before it can be easily detected, and a tumor of this size contains approximately 10^9 (1 billion) cells. For a tumor to grow from one cell to 10^9 cells requires 30 doublings, which would take five years if the doubling time were two months. After 40 doublings the tumor would weigh about one kilogram (2.2 lb), a lethal size tumor. The important point here is that a one-gram tumor with a doubling time of two months would have been present about five years and have existed for three-fourths of its life span before having reached a lethal size (exclusive of metastasis).

A more realistic view holds that tumor doubling time remains constant (i.e., relative increase in *volume* proportional to elapsed time) only up to a tumor mass of about one gram. Thereafter, the *diameter* increases in proportion to the elapsed time (see Cohen, in Schwartz, 1966). However, in a clinical setting in which we are seeking a rough estimate of the rate of tumor growth, we may assume that growth in volume remains exponential (i.e., constant doubling time up to a mass of a few grams).

Tumor doubling time can be determined by actual measurement and application of appropriate equations. Suppose a spherical tumor grows from a diameter of 2 cm to 3 cm in a period of 5 months. What is the doubling time, the time it took for the tumor to have doubled in volume? Here is the procedure for solving this problem:

1. *Find change in volume at 5 months.* The volume of a sphere is proportional to the cube of the diameter. So, the ratio of the 3-cm to the 2-cm tumor volumes is $3^3/2^3 = 27/8 = 3.4$; that is, in 5 months the tumor volume has increased 3.4 times.

2. *Apply equations for doubling time.* Let n = number of doublings. Then

$$2^n = \frac{final\ volume}{initial\ volume} = \frac{(final\ diameter)^3}{(initial\ diameter)^3} \qquad (1)$$

$$doubling\ time = \frac{elasped\ time}{n} \qquad (2)$$

In our example,

$$2^n = 3.4$$

and by using logarithms,

$$n \log 2 = \log 3.4$$
$$0.3010n = 0.5315$$
$$n = 0.5315/0.3010 = 1.77\ doublings$$

From equation (2)

$$doubling\ time = 5\ months/1.77\ doublings$$
$$= 2.94\ months$$

Doubling times for malignant tumors usually range from about two to six months. However, a benign granuloma may occasionally fall within this range of doubling times.

In Figure 16.05 we see that the growth in volume of a spherical tumor increases at a much greater rate than the increase in diameter. Thus, as the diameter doubles (in this case) from one to two cm in the first nine months, the volume (or mass) increases eight

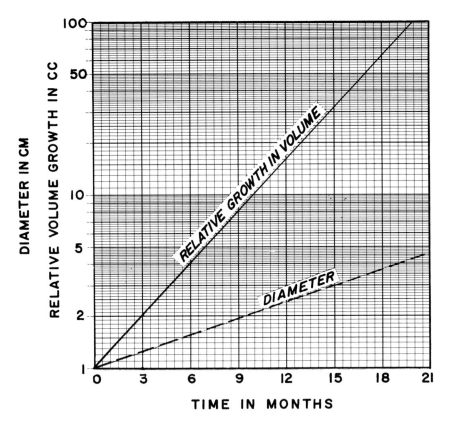

Figure 16.05. Curves showing the growth in tumor volume (assuming exponential growth) relative to the increase in diameter. This indicates that tumor growth rate is underestimated, for example, in comparing the diameter of a mass in the lung on successive radiographs, rather than the volume. Actually, growth may be exponential only up to a tumor mass of a few grams.

times during the same time span. If we were to compare the growth of a tumor, such as a bronchogenic carcinoma, on successive radiographs on the basis of the diameter we would greatly underestimate its growth rate. During the exponential growth phase of a tumor, the volume increases in proportion to the cube of the diameter. In the above example, the diameter has doubled, and so the volume increases as $2^3 = 8$ times.

In the absence of a cell loss factor (see page 75), tumor growth rate would be faster and doubling time shorter. This is the basis of *potential doubling time*. In radiotherapy, rest periods between therapy cycles may lead to a smaller loss factor and an increased doubling time, resulting in little or no regression of the tumor.

Response of Neoplasms to Irradiation

During a course of radiotherapy, a favorably responding tumor undergoes progressive loss of volume. Dead cells—those killed by radiation as well as those dying from old age—may simply liquefy by action of enzymes, or be engulfed and digested by macrophages (see pages 131–133). Of course, cure means complete disappearance of the neoplasm grossly and microscopically, and subsequent absence of local recurrence or metastasis.

What are the factors that determine tumor response to irradiation? We shall limit our discussion to the most widely used radiation (x and gamma rays) that releases electrons in tissues as described on pages 18–26. These primary electrons, in turn, cause ionization and excitation of atoms and molecules along their paths. The resulting energy deposition causes radiolysis of water with release of free radicals and hydrated electrons, which finally produce the radiobiologic lesion (see pages 58–59).

Because of the increasing experience with high-LET radiation in certain radiotherapy centers, response to this type of radiation will also be considered.

Seven factors govern the response of tumors to ionizing radiation: radiosensitivity, oxygen effect, fractionation of dosage, volume effect, radiation quality (LET), dose-rate effect, and abscopal effect. Each will receive attention under a separate heading, although they are all interrelated.

Radiosensitivity

This subject is complicated by the fact that results obtained from experiments with cell cultures do not always have a direct bearing on radiotherapy. Still, they have enlightened our understanding of cellular response at least under controlled laboratory conditions. In an attempt to simplify the subject and present the more pertinent data, we shall deal separately with radiosensitivity as it applies to cell cultures and to tumors.

Cellular Radiosensitivity (Inherent). As we have explained earlier (see pages 101–104), cellular radiosensitivity studies involve the exposure of cell cultures to a series of different radiation doses, from which cell survival curves are plotted. For our purpose, the most important concept derived from a cell survival curve is the mean lethal dose, D_o, the dose which reduces the surviving fraction of the cell population to 37 percent between two points along the straight portion of the curve. Note that D_o represents an inverse relationship to radiosensitivity: the more radiosensitive the cells, the smaller the value of D_o (i.e., it takes less radiation to reduce the cell population to a particular fraction when the cells are more highly radiosensitive). Thus, in a true sense, D_o serves as a measure of radioresistance, or 1/radiosensitivity. The D_o for almost all mammalian cells exposed to low-LET radiation (x or gamma rays) is about 130 rads $\pm$ 50%, representing a narrow range of 65 to 195 rads. Bear in mind that this yields a ratio of about 3:1 between the least and most radiosensitive cell population.

TABLE 16.03

MEDIAN LETHAL DOSES (LD$_{50}$) FOR VARIOUS ANIMAL TUMORS.*

Tumor/Mouse Strain	*Radiation* kV_p/HVL *(Cu)*	LD_{50} *In Vivo/ Fractions*	*Author*
A. *Isogenic Breast Carcinomas*:			
Isografts/dBA	200/0.9	13,000/1	Goldfeder
Isografts/C3H	240/0.4	5,700/1	Cohen
Isografts/C3H	250/2.0	5,000/1	Suit
Spontaneous/C3H	250/2.0	6,200/1	Suit
Spontaneous/C3H	200/0.9	9,900	DuSault
B. *Homologous Transplanted Mouse Tumors*:			
Sarcoma (MeCh)/C3H	240/0.4	2,500	Cohen
Adenocarcinoma/C3H	200/1.0	11,400	Ting
C. *Rat Tumors*:			
Flexner-Jobling	200/—	2,600	Melnick

*From data compiled by L. Cohen in Schwartz, E.E. (Eds.). *The Biological Basis of Radiation Therapy*, 1966. Courtesy of J. B. Lippincott.

Tumor Radiosensitivity. Intact tumors within the body present a much more complicated situation than a homogeneous cell culture. In fact, many tumors have been found to contain cell subpopulations having different degrees of radiosensitivity. Therefore, a different criterion of radiosensitivity must be applied. In experimental tumors in small animals the concept of *median lethal dose, LD_{50},* has been introduced as a measure of tumor responsiveness. *The LD_{50} is defined as the dose that will cure 50 percent of tumors under specified conditions.* Obviously, we can determine LD_{50} much more readily in the laboratory than in medical practice. In fact, L. Cohen has compiled a table listing the LD_{50} for 27 varieties of animal tumors (see Table 16.03). You can see that the LD_{50} values range from 2500 to 13,000 rads, but two-thirds lie within the limits of 2500 to 8000 rads for a ratio of about 3:1; this approaches the ratio of maximum to minimum radiosensitivity in cell cultures as mentioned in the preceding section. We must hasten to say that not only do the D_o and LD_{50} represent different standards of radiosensitivity, but that experimental conditions are far different in cell culture and in intact tumors. Still, the overall range of radiosensitivity is of the same order of magnitude in both instances.

In human irradiation therapy we find that the ratio of doses needed to arrest the most radiosensitive tumors is also about 3:1. However, we are dealing here with fractionated dosage, which is not strictly comparable to the single dose condition in the laboratory. Again we find a narrow range of radiosensitivity for various kinds of tumor populations.

Some years ago Shields Warren proposed the following guide to radioresponsiveness of human tumors, based on curative doses:

1. Radiosensitive—less that 2500 rads.
2. Radioresponsive—2500 to 5000 rads.
3. Radioresistant—more than 5000 rads.

However, just because a tumor is "radioresistant" does not necessarily mean that it cannot be cured by irradiation; it simply means that a larger dose and a longer response time are needed for cure. For example, squamous cell carcinoma of the skin falls into the so-called radioresistant category, yet it can often be cured by doses of 6000 to 7000 rads (60 to 70 Gy), although response time is slower than with radiosensitive tumors. Therefore, the concept of radioresistance is not absolute and so must be qualified. The important point is that the curative dose should ideally give a high probability of local tumor control without exceeding an acceptable level of injury to normal tissues in the radiation field. In this regard, we must note that the tolerance of certain vital structures (e.g., liver, kidneys) is much lower than that for normal skin.

On the other hand, malignant lymphomas of the non-Hodgkin type respond well to doses within the radioresponsive range (about 4000 rads in 4 weeks). Yet, they are seldom cured by irradiation alone due to their tendency to widespread noncontiguous involvement of lymph nodes and extranodal sites.

The rapidity of tumor shrinkage or involution may serve as an index of its radiosensitivity. Let us see what factors influence this process.

HIGH PROLIFERATIVE ACTIVITY (HIGH MITOTIC RATIO). This means high mitotic activity—frequent mitoses and long mitotic phase. Recall the Law of Bergonié and Tribondeau which states that radiosensitivity parallels mitotic activity. Tumors whose cells show high mitotic activity would be expected to shrink rapidly as the injured cells die or fail to reproduce. However, cells do not have to be in mitosis to suffer radiation injury—they also undergo radiation injury during the G_2 phase. Besides, lymphocytes unlike other cells are killed by prompt lysis following irradiation.

SHORT LIFE SPAN OF TUMOR CELLS (HIGH CELL LOSS FRACTION). Tumors that have a high cell loss fraction, that is, short average life span, will shrink faster because inactivation of cycling cells by irradiation prevents or delays replacement of lost mature cells. This exemplifies the response of a vertical or rapid renewal population (see pages 80–81) to irradiation. On the other hand, tumors composed of cells having a long life span will take longer to manifest radiation effects on cycling cells because of the relatively long survival of mature cells; such tumors will appear to be radioresistant because of slower shrinkage, but eventually may be curable if adequate dosage can be given, as is true with squamous cell carcinoma of the skin and cervix.

EDEMA AND INCREASED VASCULARITY. In some tumors, edema (swelling by accumulation of water) and increased vascularity (dilatation of blood vessels) may occur, especially in the early period of radiotherapy. This may hide the shrinkage of the tumor. As the edema and excess vascularity subside, the tumor regression becomes apparent.

It must be emphasized that therapy should not be curtailed just because a tumor shrinks rapidly. Radiotherapy aims to approach LD_{100}, that is, a dose that inactivates 100 percent of tumor cells. Discontinuing therapy prematurely would permit surviving cells to reproduce and cause tumor recurrence.

Oxygen Effect

As we have emphasized above (pages 122–126), fully oxygenated cell cultures have a radiosensitivity 2.5 to 3 times greater than anoxic (no oxygen) cultures, provided the oxygen is present at the time of irradiation. This, the *oxygen effect*, has been assumed to apply to irradiation therapy of human cancer.

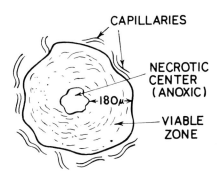

Figure 16.06. Oxygen status of a typical mammalian cancer, surrounded by capillaries that provide oxygen by supplying blood to the tumor cells. Oxygen cannot diffuse farther than about 180 microns from a capillary. Therefore, the center of a tumor larger than about 360 microns in diameter will become anoxic. This accounts for the impaired radiosensitivity of the central zones of most tumors.

We do know that mammalian tumors, when they grow to a sufficiently large size, have a deficiency of oxygen at the center (see Figure 16.06). Thomlinson and Gray (1955), found that the maximum diffusion distance for oxygen (O_2) in tissues is about 180 microns (μm). It so happens that the thickness of a tumor's outer oxygenated layer remains virtually constant—160 to 180 μm—even as the hypoxic (oxygen-poor) center enlarges. Thus, if the central tumor zone lies farther than the critical 180 μm from the peripheral capillaries, it will contain a core of necrotic (dead) cells on account of the anoxic conditions, intermingled with cells that have managed to survive the anoxia. Since these cells are 2.5 to 3 times more radioresistant than the fully oxygenated cells, they can survive and repopulate, with resulting tumor recurrence. This may well explain the failure to cure certain tumors despite apparently adequate radiation dosage.

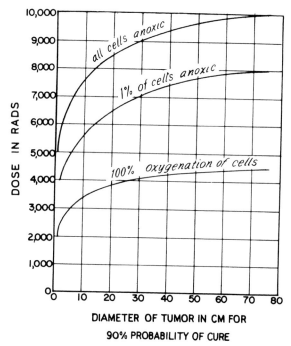

Figure 16.07. Dose required for 90 percent probability of cure, relative to the diameter of the tumor and its accompanying state of oxygenation. Note that the "curative" dose increases with the size of the tumor and the degree of anoxia.

The 90 percent curative doses in animal tumors of various sizes and degrees of oxygenation are shown in Figure 16.07, based on the classic work by H. B. Hewitt as quoted by Fowler, Morgan, and Wood (1963). Note how the curative dose increases with the tumor diameter (center presumably poorly oxygenated), and therefore with the percentage of hypoxic cells. Such experiments bear out the assumption that deficient oxygen occurs in at least some human tumors as they outgrow their blood supply, as well as under certain special circumstances such as torsion (twisting on vascular pedicle), and scarring.

We may summarize the problem of *tumor oxygenation* by stating that it is a function of four factors: (a) the distance between the most central cells in the tumor and the nearest capillaries, (b) the O_2 tension in the capillary bed, (c) the rate of utilization of O_2 by intervening viable cells, and (d) the rate of physical diffusion (spread) of O_2 into the tumor mass. We have already discussed (a) above. Obviously, the greater the O_2 tension in the capillary bed, the farther the O_2 can diffuse into the tumor. If cells between the capillary supply and the center of the tumor have a high rate of O_2 consumption, insufficient O_2 will remain to reach the center. Finally, any condition that favors the physical diffusion of O_2 through the tumor will improve oxygenation at its center.

How can we lighten or overcome the effects of anoxia in tumors so as to increase their radiosensitivity? Several methods are available, although they differ widely in effectiveness.

1. *Fractionation therapy* in which a series of daily doses is given; the advantages will be discussed in the next section. At this point we should note that fractionation results in reoxygenation of hypoxic regions of a tumor as it shrinks. However, reoxygenation is not always complete.

2. *Increased oxygen tension*, achieved by placing the patient in a hyperbaric (high pressure) chamber at two to three atmospheres of pressure. Unfortunately, the results have thus far been equivocal, but clinical trials are continuing.

3. *High-LET radiation*, which provides a lower oxygen enhancement ratio (see pages 124–125) than does photon radiation, under intensive evaluation in a number of oncology centers. Included are neutrons, negative pi mesons, and heavy charged particles. Such radiation would be expected to increase the radiosensitivity of anoxic tumor cells with resulting greater curability. However, this remains to be proved.

4. *Hyperthermia* (above normal temperature), by means of which the temperature of a tumor is increased to about 43 to 45°C (normal body temperature 37°C), which increases the radiosensitivity of anoxic cells to a greater degree than oxic cells. In a clinical trial reported by Bicher et al. (1980), a fractionated series of hyperthermia treatments induced by a microwave unit, followed by fractionated irradiation (low-LET), seemed to improve tumor response. It has been found that synergism between hyperthermia and subsequent irradiation shortly thereafter (i.e., total effect of both greater than sum of individual effects) is most evident during the ordinarily radioresistant S-phase of the cell cycle. In addition, the anoxic cells exhibit increased radiosensitivity. Moderate hyperthermia (40 to 42 C) also increases blood flow within the tumor, thereby bringing more oxygen to the anoxic regions. Because of the normally sluggish blood flow in tumors, they retain heat better than normal tissues and therefore suffer more damage. Hyperthermia induced by various modalities in association with irradiation is being actively evaluated in a number of oncology centers.

Dose Fractionation

Since the *goal* of curative radiotherapy is to deliver a lethal dose to all the cells of a particular tumor without causing irreparable damage to the neighboring normal cells, we must know about the differential response of the two kinds of tissue in question. Obviously, the greater the injury to the tumor cells, the higher will be the probability of tumor control. In radiotherapy this differential effect is embodied in the concept *therapeutic ratio*, to be discussed on pages 259–262.

The *differential response* of a tumor and its neighboring normal cells (both kinds being included in the therapy beam) depends on their *differential radiosensitivity* and their *differential recovery* rate. Both these factors strongly influence curability.

Although most normal and tumor cells have a narrow range of radiosensitivity, there is nevertheless a sufficient practical difference to be significant, at least in the case of some tumors. For example, lymphomas exhibit appreciably higher radiosensitivity than normal tissues.

Of at least equal importance is the differential recovery rate. Most normal cells show better recovery than tumor cells to a particular radiation dose. Several possible explanations have been offered. First, normal tissues harboring tumors often display less reproductive activity and are therefore less vulnerable to injury by ionizing radiation. Second, normal tissues have a better vascular and nerve supply, which helps maintain more adequate nutrition. When normal cells experience nuclear damage, repopulation occurs more rapidly from nearby undamaged cells because of the ensuing reduction in cell time. Finally, whereas various mammalian cells show narrow differences in radiosensitivity in the straight portion of the survival curves, various cell lines show wider differences in radiosensitivity in the shoulder portion. Since fractionation therapy involves daily return of shoulders, even small differences in the ability of normal and tumor cells to recover are magnified during such a course of therapy (see Figure 16.08).

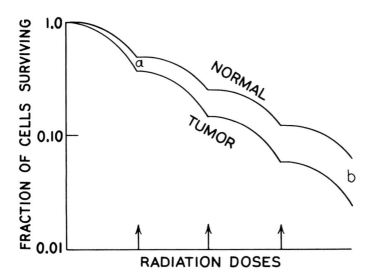

Figure 16.08. Increasing differential survival with number of daily fractions. At b, after four dose fractions, there is greater differential survival between the normal tissues and the tumor than at a after one fraction. Thus, small differences in survival are amplified by fractionation. *(Adapted from Duncan W and Nias AHW, 1977.)*

Whenever possible, full advantage should be taken of the differential radiosensitivity and recovery rates between the tumor and its associated normal tissue. Any factor that enhances these differences in favor of normal tissue should improve the chances for tumor arrest. *Dosage fractionation is just such a factor.* It has been known since 1918 that treatment given in a series of small doses of x or gamma rays requires a larger total dose and is better tolerated than a single-shot treatment, for the same tumor cure rate. Such division of total treatment dose into daily increments is called *fractionation.* Coutard (1932) firmly established the principle of fractionation therapy in head and neck cancer, especially the larynx, ascribing the markedly improved results to the following:

a. Better recovery of normal tissues as compared with cancer, that is, enhanced differential recovery with an improved therapeutic ratio.

b. Greater opportunity to irradiate the tumor during periods of heightened radiosensitivity, that is, during mitosis (also, during G_2 part of interphase as we have learned since his time).

A classic in clinical research is Strandqvist's report in 1944 on x-ray treatment of nearly 300 assorted skin and lip cancers ranging in size from five to 30 cm^2, with a single-dose and with various fractionated-dosage schedules; the smaller the daily dose, the greater the total dose needed for cure. Plotting the data on log-log paper, he fitted the resulting straight line curve between overdosage (necrosis) and underdosage (recurrence). This type of curve, known as a time-dose *regression curve* or *isoeffect curve*, is shown in Figure 16.09. Sample exposures expressed in R (as in his original data) appear in Table 16.04. A number of reports since Strandqvist's time have essentially verified his observations. However, his dosage schedule must not be followed blindly because skin necrosis can still occur in some cases. Furthermore, the isoeffect curve is not necessarily a straight line, for it may actually represent an oversimplification.

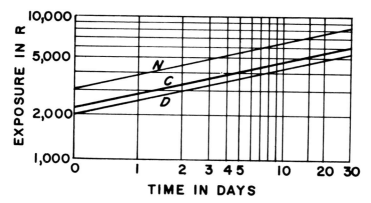

Figure 16.09. Time-dose "isoeffect" curves for skin, representing the total exposures for various daily fractionation schedules (time in days) for cure of skin cancer (curve *C*), skin necrosis (curve *N*), and skin desquamative reaction (curve *D*). These are straight lines on a log-log plot, and have the same slope, *n* = 0.22. At 0 on the time axis is the single exposure needed for each particular effect. Actually, these curves were derived by best fit of the clinical data; there is a degree of variability in the required exposures from patient to patient. (*After Strandqvist. Acta Radiol [Suppl 55], 1944.*)

Assuming the validity of the straight line character of the isoeffect curve in most cases, we can express it in the form

$$D = kt^n \qquad (3)$$

where D is the dose for a specific effect in a particular tissue, k is a constant, t is the overall treatment time in days, and n is a constant (always a positive fraction) called the *recovery exponent*. (Note that n here is not the same as the extrapolation number in a cell survival curve.)

The meaning of k and n requires comment. If a specific effect such as tumor lethality is to be the goal of a *single* irradiation, $t = 1$ day and $t^n = 1$, regardless of the value of n. In this case, $D = k$, so we may define k as the single dose of radiation that will produce the desired effect (tumor lethality in this instance). Table 16.05 gives estimates of the comparative single doses for various effects in normal skin and representative cancers, with 200-kV and megavoltage radiation.

TABLE 16.04

TOTAL EXPOSURES WITH VARIOUS FRACTIONATION SCHEDULES
TO ACHIEVE A 90 PERCENT CURE RATE (LD_{90})
OF SMALL SKIN CANCERS.*

Daily Fractions	Total LD_{90}
	R
1	2250
4	3400
7	4100
21	5000
30	5400

*Data from Strandqvist M. *Acta Radiol (Suppl 55)*, 1944.

TABLE 16.05

ESTIMATES OF RADIOSENSITIVITY OF HUMAN
SKIN AND SOME TYPICAL NEOPLASMS.**

		10-cm Diameter Field — Single Dose	
		Dose at 200 kV	Dose at 1 MeV and above
NORMAL SKIN		rads	rads
threshold erythema dose		750	1000
dry desquamation, median effective dose		1500	2000
exudative reaction, maximum tolerance dose		3000	4000
NEOPLASMS			
epidermoid carcinoma	LD_{50}	2000	2600
	LD_{90}	2500	3300
breast carcinoma	LD_{50}	1300	1700
	LD_{90}	1600	2100
lymphoma	LD_{50}	900	1200
	LD_{90}	1100	1400

**From data of L. Cohen in Schwartz, E. E. (Ed.): *The Biological Basis of Radiation Therapy*, 1966. Courtesy of J. B. Lippincott.

Note that the above data are for single doses and tumors no greater than 10 cm in diameter. Also, the "curative" dose for breast carcinoma approaches that for epidermoid carcinoma when prolonged fractionation is used; thus, the LD_{90} for both types of tumor is about 5000 rads with 26 daily fractions.

The constant n is the slope of the time-dose regression line: the larger the value of n, the steeper the slope and the greater the recovery rate of irradiated tissue.

Strandqvist's curves give values of k = 1500 R for a normal skin reaction (i.e., dry desquamation), and 2000 R for cure of skin cancer. The slopes of these curves are identical with an n of about 0.22, indicating equal recovery rates of normal and cancerous skin.

However, a number of authorities have found that the curves do not have the same slope. L. Cohen (1968) analyzed numerous published articles and plotted the composite data as shown in Figure 16.10. He concluded that reasonable values may be n = 0.33 for normal skin, and 0.22 for skin cancer, indicating a greater recovery rate for normal skin. These data are borne out by clinical experience, despite the extremely small difference in the slopes.

The preceding discussion has dealt with empirical curves obtained by plotting and smoothing clinical data derived from conventional daily fractionation schedules. However, we cannot be certain that other fractionation schemes might not produce better tumor control with less injury to normal tissues. One modifica-

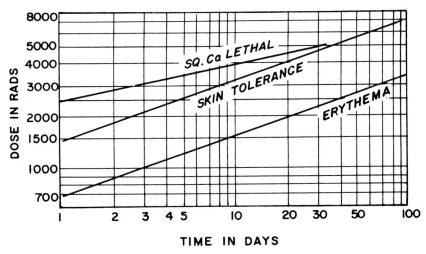

Figure 16.10. Composite time-dose "isoeffect" curves derived by L. Cohen from five published sources. These show a steeper slope (n = 0.33) for skin tolerance than for cure of skin cancer (n = 0.22), indicating a slightly greater recovery rate for normal skin. Data have been normalized to a 10-cm or equivalent field. (*Adapted from Cohen L. Brit J Radiol 41:522, 1968.*)

tion is the so-called *split course* in which the daily-fractionated therapy course is interrupted by a two- or three-week rest period after one-half the total dose has been delivered. In some patients, especially those in poor general condition harboring advanced lesions, such a split course often allows recovery of normal tissues and improvement in the patient's health status during the rest period. Whether split course therapy gives as high a control rate as ordinary continuous fractionation (daily treatments Monday through Friday, Saturday and Sunday off, no rest period) has not been definitely established but it does improve the patient's tolerance locally and generally.

In 1965 Ellis introduced a unit of dosage that provides a basis of comparison among different fractionation schedules and permits variation in them while retaining equivalent biologic effects. This unit is the *ret* (*r*ad equivalent *t*herapy), the biologic dose being called the *nominal standard dose*. To arrive at its formulation he made certain assumptions based on data acquired from radiobiologic research and clinical experience. These include the following:

1. *Normal and malignant cells respond similarly* to radiation insofar as lethality, loss of reproductive capacity, and intracellular repair of sublethal injury are concerned. This is supported by the similarity of survival curves of various cell cultures.

2. *Only normal cells can recover* by intervention of extracellular or homeostatic factors such as immune factors, new capillary formation, and repair of connective tissue by fibroblasts. *Malignant cells* are not subject to such extracellular repair processes, this being a major difference in the response of normal and malignant cells. (However, some cancers produce a substance called *angiogenesis factor*, which promotes ingrowth of new capillaries.)

3. *Intracellular repair is rapid* (Elkind type), occurring within a few hours, whereas *extracellular repair is slow*, requiring days or weeks. Hence, intracellular recovery depends on the number of fractions, while extracellular recovery depends on the overall time in which treatment is given.

On the basis of such considerations Ellis separated the *fractionation factor* (related to fast, intracellular recovery) from the *total treatment time* in days (related to slow recovery by repopulation).

Since skin cancer can undergo *only intracellular repair* (if it occurs at all), Ellis adopted the value 0.24 as the recovery exponent of the isoeffect curve for *rapid recovery*. (The slope according to Figures 16.09 and 16.10 is actually 0.22, but he modified it to 0.24 for five-day per week treatment.) At the same time, the isoeffect (time-dose regression) curve for normal skin contains two components— one for fast recovery and the other for slow. Since the isoeffect curve for normal skin has a slope of 0.33, the difference between the exponents (0.33 − 0.22 = 0.11) represents the exponent for *slow recovery* related to the overall treatment time in days. From this reasoning came the following equation:

$$D = NSD \times N^{0.24} \times T^{0.11} \ rads \qquad (4)$$

where D is the total dose in rads (for megavoltage photons), NSD is the nominal standard dose in rets, N is the number of fractions, and T is the total treatment time in days between the first and last treatment. Ellis emphasizes that the fractions must be separated by at least 16 hours and must be evenly spaced (5-day per week schedule), and T must be in the range of 3 to 100 days.

The NSD is a dose representing a biologic effect related to normal connective tissue tolerance as observed in pig skin exposed to ionizing radiation. Thus, it is the real limiting factor in radiotherapy. Furthermore, it is incorrect to regard the NSD as the single dose that would have the same effect as a fractionated series, because the factor of slow recovery would not have time to come into play and equation (4) would not be valid.

Although Ellis regards 1800 rets as the maximum normal connective tissue tolerance, he recommends that each radiotherapy center establish its own value depending on local experience. In the case of lymphomas, which are highly radiosensitive and in which it is unlikely that the required dose would reach normal tissue tolerance, only $N^{0.24}$ need be included in equation (4). So, for the malignant lymphomas we can obtain a *tumor standard dose in rets* according to

$$D = TSD = N^{0.24} \qquad (5)$$

Ellis warns that in applying equation (4) to a portion of the treatment course the dosage in rets applies only to the portion of

Elements of Radiobiology

the fractions delivered. For example, if the total planned *NSD* is 1800 rets in 30 fractions, and only 15 fractions are given, then the given *NSD* is 15/30 × 1800 = 900 rets.

Radiobiologists disagree as to the universal validity of the *NSD* concept. While it is not a substitute for clinical experience and careful planning, it does provide an approximate common denominator for intercomparison of various treatment schedules and for introducing a degree of flexibility in varying treatment dosage schemes. For example, 400 rads twice weekly has about the same radiobiologic effect as 200 rads five times a week. However, most radiotherapists use a treatment schedule approximating the 5-day per week plan in routine therapy because of past experience, and they are reluctant to depart from it until a better fractionation schedule can be conclusively demonstrated.

How can we explain the superiority of fractionated therapy over single-dose therapy? The experimental and scientific foundations have already been described, but these will now be applied to radiotherapy. Observations *in vitro* (cell culture) and *in vivo* (live animals) has disclosed the operation of four major radiobiologic factors in fractionation therapy; these have been called *the four R's*:

1. Repair of sublethal injury (Elkind or fast recovery),
2. Reoxygenation
3. Redistribution of cells in the cell cycle
4. Repopulation (slow recovery)

These will now be discussed briefly.

Repair of Sublethal Injury. This is shown experimentally in the reappearance of the shoulder in the cell survival curve (see page 114) within a few hours after each successive dose. First published by Elkind (1959), this effect has been designated *fast* or *Elkind recovery*. As the number of dose fractions is increased during a particular overall treatment time lasting several weeks, the straight portion of the survival curve becomes less steep (i.e., smaller slope, larger D_o) indicating lessened radiosensitivity (see Figure 16.11). If such diminished radiosensitivity should be the same in both the tumor and normal cell populations there would be no advantage in fractionation over single-dose therapy. Actually, while

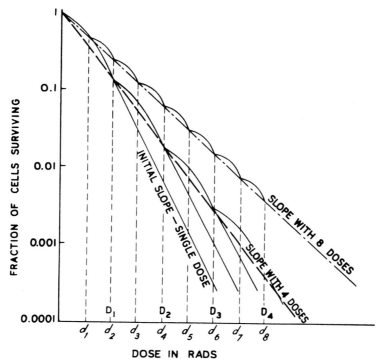

Figure 16.11. Modification of cell survival curves by splitting a single dose into multiple doses. Separation of doses by a time gap of a few hours results in reappearance of the shoulder (sometimes referred to as "return of *n*"). The larger the total number of fractions for the same total dose, the smaller the slope of the cell survival curve.

most oxygenated normal and tumor cells have D_0 values within a range of only about 0.65 to 1.95 Gy (65 to 195 rads), the D_q values (width of shoulder) have a wider range—1.0 to 5.5 Gy (100 to 550 rads) in mice (Duncan and Nias). This suggests at least the possibility of a difference in capacity for repair of sublethal injury among various cell populations. However, the oxygen effect is believed to play an even greater role in fractionation therapy.

Cell "killing" in the shoulder portion of the cell survival curve results from single-hit events rather than from accumulation of sublethal injury (see page 104). Therefore, during the usual course of fractionation therapy with 200-rad fractions, virtually all cell killing is due to single-hit events.

Reoxygenation. As we have already seen earlier in this chapter (see pages 241–243) anoxic cells are about one-third as radiosensitive as fully oxygenated cells according to their respective D_0 values. This holds true for many different kinds of normal and tumor cells. Since about 10 to 20 percent of cells in most tumors are likely to be hypoxic (oxygen-deficient) while normal cells are usually well-oxygenated, additional oxygen should have a relatively greater sensitizing effect on tumor than on normal tissue. Thus, oxygen improves the therapeutic ratio.

We should expect that during a course of fractionated therapy, oxygenation of the tumor would improve for these reasons:

1. Oxygenated cells are killed first.
2. Tumor shrinks.
3. Hypoxic cells now have more oxygen available because more blood vessels are closer to them, and there are fewer cells competing for the available oxygen.
4. Radiation also improves oxygenation by decreasing edema, thereby alleviating stagnant anoxia.

In summary, then, dosage fractionation in radiotherapy should lead to progressively improved oxygenation of hypoxic tumor regions, thereby increasing overall radiosensitivity. Fractionation enhances the therapeutic ratio through the oxygen effect.

Repopulation. It would seem highly desirable that during a course of radiotherapy normal cells proliferate rapidly to maintain their normal state, while tumor cells do not. Although experimental evidence is not entirely conclusive, it does suggest that after irradiation normal cells undergo a shortening of cell cycle time, whereas tumor cells of the same type show an opposite effect. In other words, radiation causes normal cells to display an increased proliferation rate, and tumor cells a decreased proliferation rate. Actually, repair of sublethal injury and oxygen effect may be much more important than repopulation during fractionation therapy.

Redistribution of Cells in the Cell Cycle. As we have already seen (pages 120–121), radiation tends to synchronize cells in their cycle for one or possibly two mitoses. It would therefore be anticipated that during fractionation therapy the cells that happen to be in a sensitive phase such as G_2 or M would be killed first, leaving an increased fraction of resistant cells. The effect of subsequent irradiation would then depend on the time between successive doses, and the size of the dose. As it turns out, the doses ordinarily used in radiotherapy are not large enough to induce appreciable synchrony, so this factor seems to be inconsequential in depleting tumor cells relative to normal cells in clinical practice.

Volume Effect

The irradiated volume must contain the entire tumor and an estimated surrounding layer of apparently normal tissue. However, this *volume of interest* should be kept as small as possible for two reasons. First, normal tissue tolerance varies inversely with the irradiated volume; that is, the larger the volume the smaller the tolerance. Involving such factors as the release of toxic products, inactivation of normal cells with impairment of tissue repopulation, injury to vascular connective tissue, and changes in host immunologic response, the volume effect plays a major role in radiotherapy. Because a large tumor volume requires inclusion of a correspondingly large volume of normal tissue, there is greater risk of irreparable damage than with a small tumor carried to the same total dose. Furthermore, a larger tumor requires a larger curative dose, owing to the fractional nature of cell killing in a cell population (see page 105), as well as the oxygen effect. Thus, because of its poorer blood supply centrally, a larger tumor is more apt to contain radioresistant hypoxic cells. On the other hand, about 90 percent of subclinical *carcinoma* metastasis to lymph nodes are controlled by 5000 rads/5 weeks (Fletcher), while the primary tumor usually requires a dose in the range of 6500 to 7000 rads.

Second, the *integral dose* increases with an increase in the irradiated volume. Integral dose is a mathematical concept, first introduced by Mayneord, specifying how the absorbed dose may be computed for the *entire irradiated volume*, from the skin of the entrance port to the skin of the exit port. For example, if a mass of tissue weighing 100 grams undergoes irradiation to a uniform absorbed dose of 200 rads, the integral dose would be

$$100 \ grams \times 200 \ rads = 20,000 \ gram\text{-}rads$$

In actual practice, the computation is not this simple because the beam is attenuated as it passes through the body, and a uniform absorbed dose is not achieved. At any rate, the integral dose concept agrees with clinical experience, namely, that large tissue volumes do not tolerate a particular absorbed dose as well as do smaller volumes. In practice, integral dose rarely enters directly into radiotherapy planning because dosage schedules for various tissue volumes have evolved over many years on the basis of clinical experience. Furthermore, the integral dose concept ignores the importance of the nature of the irradiated tissue; for example, irradiation of the liver obviously has a much more detrimental effect than irradiation of a like volume of pelvic tissue to the same total dose.

Radiation Quality

At present, megavoltage x rays generated by linear accelerators are becoming the mainstay of irradiation therapy. Commonly used are x rays with an energy of 6 MV, although higher energy linacs (e.g., 20–25 MV) are also available, particularly for electron beam application. Cobalt beam units are gradually becoming obsolete because of the continual decay, with corresponding adjustment of treatment times, as well as the large penumbra and limited field size. Although megavoltage radiation (one million volts and higher) has brought about a marked improvement in dosage distribution, with skin and bone sparing, energy deposit in tissue still occurs via the release of energetic electrons, which are in the low-LET range, just as are those produced by 250-kV x rays. In fact, megavoltage radiation has a relative biologic effectiveness (RBE) 15 to 20 percent *less* than that of orthovoltage x rays.

Low-LET radiation (usually fast electrons released by x or gamma rays) in low dosage may not succeed in inactivating multiple targets in the cell to engender a lethal effect. Hence, only sublethal effects may result at low dosage; recovery is manifested by the shoulder in the cell survival curve. With sufficiently large doses of low-LET radiation, multiple targets may be hit, increasing the probability of cell death (i.e., loss of reproductive ability). Intermediate-LET radiation such as fast neutrons deposit energy by releasing recoil or "knock-on" protons which, in turn, produce ionization with clusters of closely spaced ions; even at low dose levels these have a much greater likelihood than electrons of hitting multiple targets and causing irreversible cellular damage.

We must not lose sight of the *oxygen effect in relation to LET*. As we have already noted, the ability of low-LET radiation to inactivate cells depends strongly on the presence of oxygen, the oxygen enhancement ratio (OER, see pages 124–126) being about 2.5. Intermediate-LET radiation such as fast neutrons has an OER of 1.6, and high-LET radiation such as 2.5-MeV alpha particles an OER of 1.3. Thus, as LET increases, the oxygen effect approaches unity; that is, it nearly vanishes, thereby improving the likelihood of inactivating anoxic cells.

In the case of *pions* (see pages 277–278) their "star bursts" within a tumor should represent very high-LET radiation. However, contamination of the pion beam with other radiation such as fast neutrons released during interaction with tissue atoms raises the average OER to a range of 1.6 to 1.8—higher than for a pure fast neutron beam. The practical advantages of pion beam therapy have still to be determined by clinical trial.

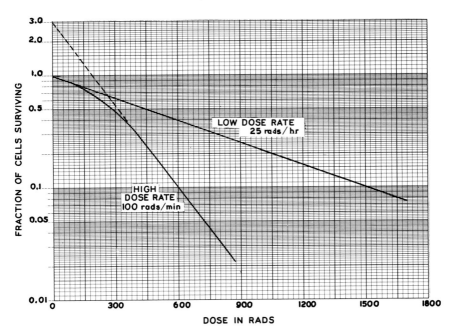

Figure 16.12. Effect of dose rate on survival of hamster ovarian cells. These show greater radiosensitivity (steeper slope) at 100 rads *per minute* than at 25 rads *per hour*, at a temperature of 20 C. *(After Duncan W and Nias AHW, 1977).*

Dose-Rate Effect

Radiotherapy of human tumors has been generally in the dose-rate range of about 100 rads (1 Gy) per min to several rads per min. Under these conditions no significant dose-rate effect appears; that is, the cell-killing potential for equal total doses and a similar fractionation schedule is virtually the same throughout this dose-rate range. However, as the dose rate with x or gamma rays is reduced to very lower levels (i.e., marked protraction), a definite dose-rate effect becomes apparent. In Figure 16.12 we see the survival curves for Chinese hamster cells *in vitro*, exposed to cobalt 60 gamma rays at various dose rates. Note the decreasing slope of the curves (i.e., lowered radiosensitivity) as the dose rate is decreased from about 100 rads (1 Gy) per min to 0.36 rad (0.0036 Gy) per min. There is a dose-rate effect even between about 100 rads per min and 36 rads per min with these particular cells. This raises the possibility that therapy with a low-output cobalt 60 source could result in undertreatment of a tumor as compared with one having a normal output. It should be pointed out, however, that normal cells and tumors vary in their susceptibility to dose-rate effects; that is, some animal tumors display pronounced dose-rate effects, whereas others do not. In any case, experiments with cell cultures have shown no appreciable dose-rate effects above 100 rads (1 Gy) per min.

In attempting to explain the influence of dose rate on radiosensitivity we must understand that this depends on two distinct radiobiologic processes: repair of sublethal radiation injury and cell proliferation. We have seen earlier (pages 252–253) that fractionated dosage schedules yield a less steep curve, this effect becoming more pronounced as the degree of fractionation is increased (i.e., as the number of daily fractions is increased) for the same total dose. The explanation lies in the repair of sublethal injury between fractions. Low dose-rate irradiation may be regarded as extreme fractionation, permitting repair of sublethal injury during the prolonged irradiation. With still lower dose rates, another process comes into play—cell proliferation during the markedly protracted irradiation —so that the survival curve becomes even flatter. Thus, as the dose rate is decreased to very low levels the cell system exhibits a decrease in overall radio-sensitivity.

In practice, interstitial and intracavitary therapy with radium and other suitable radionuclides represents a low-dose situation. For example, the standard treatment technic with so-called low intensity radium needles specifies a dose of 6000 rads (60 Gy) in one week (168 hours), or about 36 rads (0.36 Gy) per *hr*. This is considerably less than the usual dose rate of about 100 rads (1 Gy) or more per *min*, and may account for the distinct impression of experienced radiotherapists that continuous tumor irradiation at low dose rates may be preferable to therapy at high dose rates insofar as normal tissue tolerance and tumor control are concerned.

Abscopal Effect (Effect at a Distance)

It has been observed that irradiation of certain tumors may occasionally cause shrinkage or disappearance of remote metastatic deposits. For example, in patients with Hodgkin's disease, irradiation of a group of nodes in the neck may cause temporary shrinkage of nodes elsewhere in the body. We do not know whether this results from as yet unidentified toxic substances released in the irradiated volume, or from some other agency. Such a response is not limited to irradiation, but may follow an infection, transfusion reaction, or, rarely, biopsy of one of the lesions. In any case, this temporary effect should not be depended on for tumor control.

Therapeutic Ratio

Developed by Paterson, the concept *therapeutic ratio* serves as a useful index of local tumor radiocurability. It embodies the goal of curative therapy, which is to deliver a cancericidal (i.e., cancer "killing") dose of radiation without causing irreparable damage to the normal tissues within the treatment field. The therapeutic ratio expresses the relationship of normal tissue tolerance to tumor lethal dose in a particular clinical situation:

$$therapeutic\ ratio = \frac{normal\ tissue\ tolerance\ dose}{tumor\ lethal\ dose} \qquad (6)$$

Obviously, the greater the tolerance of normal tissue (larger numerator) and the smaller the dose needed to destroy the tumor (smaller denominator), the larger will be the therapeutic ratio and the better the chance of eradicating the tumor without excessive damage to normal tissue. Accordingly, any factor which augments the numerator or diminishes the denominator will improve the therapeutic ratio. The factors we have discussed pertaining to tumor response should affect the therapeutic ratio; they are presented in the following outline where ↑ indicates enhancement of therapeutic ratio, and ↓ the opposite.

Therapeutic Ratio Influenced by

1. *Tumor radiosensitivity*
    a. *Intrinsic*—cells poorly differentiated (i.e., high mitotic rate, long mitotic phase) ↑
    b. *Extrinsic*
        i. Location in a vascular bed ↑
        ii. Oxygen tension high ↑
        iii. Tumor volume small ↑
    c. *Acquired radioresistance* e.g., previous irradiation probably impairing blood supply, and by natural selection of more resistant tumor cells as more sensitive ones are killed off. ↓
2. *Normal Tissue Tolerance*
    a. Cell type relatively radioresistant (?) ↑
    b. Vascularity adequate (e.g., for repopulation) ↑
    c. Systemic tolerance (i.e., of body as a whole)
        i. Age—young ↑
        ii. Infection ↓
        iii. General state of health good ↑
        iv. Host resistance (immunity) ↑
    d. Volume small ↑
3. *Fractionation* (with low-LET radiation) ↑
4. *Overall treatment time long* (within limits) ↑
5. *High-LET radiation* (low OER) ↑
6. *RBE* (if greater for tumor than for normal tissues) ↑

Note that the therapeutic ratio serves only as a guide to therapy and lacks absolute significance because no fixed values can be assigned to the governing factors. Besides, normal tissue tolerance is only relative. For example, a 2 percent incidence of normal tissue necrosis may be acceptable in an attempt to eradicate a potentially curable tumor. If one were to plan radiotherapy to avoid normal tissue necrosis altogether, then some potentially curable tumors would not be cured. Thus, one has to weigh carefully risk *versus* benefit factors. Furthermore, patients must be individualized; thus, in an elderly or debilitated patient whose malignant tumor has a natural history longer than the patient's life expectancy, the dose may be kept below the level that might cause severe complications, despite the fact that the tumor may not be permanently arrested.

The therapeutic ratio may also be expressed as the ratio of t^n of normal tissue to that of the tumor as derived from the equation

$$D = kt^n$$

Accordingly, the therapeutic ratio for skin cancer (see pages 246–249) may be stated as

$$T.R. = t^{0.32}/t^{0.22} = t^{0.10}$$

From this we see that as the overall treatment time t is increased under these specific conditions, the value of $t^{0.10}$ increases, and so does the therapeutic ratio.

We must recognize that there is no single dose that "cures" all neoplasms of a particular type. In general, as the absorbed dose is increased, the probability of tumor control increases. The curve representing percent lethality as a function of absorbed dose (i.e., tumor response curve) is sigmoid or S-shaped (see Figure 16.13). So is the curve for the incidence of necrosis as a function of absorbed dose. Obviously, any factor which separates these two curves, moving the tumor response curve to the left and the necrosis-incidence curve to the right, will increase the therapeutic ratio. This important concept must be kept in mind by the therapist in planning a course of irradiation therapy.

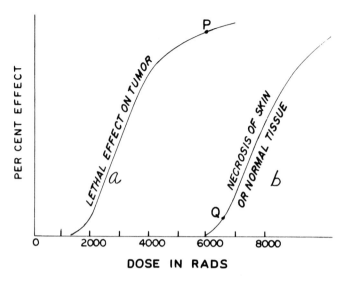

Figure 16.13. Typical sigmoid dose-response curves showing percent effect from various doses of radiation. Beyond some point *P* on curve *a*, virtually all tumor cells are killed; thus, *P* may be called the *tumor lethal dose*. At some point *Q* on curve *b* there is a steep rise in the incidence of necrosis with increasing dose. The ratio of dose *Q* to dose *P* is the *therapeutic ratio*, which becomes larger as the separation between the curves increases. (*Adapted from Paterson R, The Treatment of Malignant Disease by Radium and X rays, 1963.*)

Chapter XVII

MODALITIES IN RADIOTHERAPY

History

The earliest application of x rays to malignant disease was Emil Grubbe's treatment of breast cancer in 1896, just a few months after their discovery by Roentgen. Four years later William Snow used x rays in treating facial skin cancer. In 1903 a Chicago surgeon, Nicholas Senn, reported his experience with x-ray therapy of leukemia.

It must be emphasized that the early radiation therapists practiced single-dose application. In the absence of accurate standardization, exposure had to be determined by trial and error. Although Villard had suggested ionization of air as a base for measuring radiation as early as 1908, it was not until 1928 that an international assembly of physicists adopted this principle in defining the roentgen as the unit of radiation exposure. The roentgen was last redefined in 1962.

A great leap forward took place in the early 1920s when Lacassagne, Regaud, Coutard, and Hauthut discovered and developed the principle of *fractionation therapy*, that is, the delivery of radiation in daily doses over a period of weeks rather than as a single large dose. They found that fractionation could destroy certain malignant tumors while causing less severe changes in the normal tissues. Coutard, in particular, applied this principle to the treatment of cancers of the larynx and pharynx with marked improvement in results, thereby laying the groundwork for modern radiotherapy.

In the early 1920s orthovoltage x-ray machines operating in the 200 to 300 kV range became available and served as the backbone of radiation therapy for about 30 years, next being superseded by cobalt teletherapy. The use of cobalt 60 in teletherapy was first suggested by J. S. Mitchell in 1946 in England, but L. G. Grimmett, who left England to become chief physicist at the M. D. Anderson Hospital and Tumor Institute in Houston, Texas, actually designed the first telecobalt unit which was put into service in 1953.

Megavoltage (million volts or more) *x-ray therapy* units include the *Van de Graaff electrostatic generator* (first used in therapy in 1937); the General Electric 1000 kV$_p$ *Maxitron* (1939); the *betatron* (invented by Kerst in 1940, first used in therapy in the 1950s); and the *linear accelerator* or *linac* (first developed in this country by Varian Associates). The linac is rapidly replacing telecobalt as the mainstay of radiotherapy, and the betatron still finds application in special situations, while the Van de Graaff and Maxitron have become obsolete.

In recent years new types of radiation for beam therapy, namely, "heavy" particles such as *neutrons* and *negative pi mesons* (pions) have become available and are being investigated in several centers.

Radium first came into medical use in 1901 in Paris when Danlos and Block applied it in the treatment of lupus (a non-cancerous skin disease). Following Alexander Graham Bell's suggestion in 1903, radium was placed in hollow needles, which could be introduced directly into an accessible malignant tumor. Two years later, Robert Abbe used radium-filled tubes as intracavitary implants for the treatment of malignant tumors. In 1915 Henry H. Janeway first buried glass "seeds" containing radon gas within a tumor, following an earlier suggestion by W. Duane. Janeway was also the first radiotherapist to use a radium pack (1917) which he placed in contact with the skin to treat an underlying tumor. Later, radium teletherapy units were designed but were limited to only a few therapy centers; experience with this modality subsequently proved its value in therapy with telecobalt beams.

Sources of Radiation for Therapy

Radiotherapy may be generally classified in two main categories: (1) *teletherapy* (*tele*, far) or external radiation whose source is located at some distance from the body, and (2) *brachytherapy* (*brachy*, short) wherein the source is placed close to or actually within a tumor.

1. **Teletherapy Equipment**. Equipment for external beam therapy includes the following:
   a. *X-ray Machines*
      i. Superficial—operating in the low kilovoltage range—50–120 kV.
      ii. Orthovoltage—operating in the medium kilovoltage range—140–300 kV.
      iii. Megavoltage—operating at 1 million or more volts.
         (a) Linear accelerator (linac), 6 mV, 20 mV, or 25 mV.
         (b) Betatron, 22 to 24 mV.
   b. *Radionuclides—Gamma-ray Beams*
      i. Cobalt 60, average energy 1.25 MeV
      ii. Cesium 137, 0.66 MeV.
   c. *Heavy Particles*
      i. Neutrons
      ii. Pi mesons (pions)
2. **Brachytherapy**—radium or its substitutes.
   a. *Interstitial*—needles or seeds in tissues
   b. *Intracavitary*—tubes in body cavities

TELETHERAPY EQUIPMENT

In this section we shall describe briefly the various types of equipment that are available for teletherapy. They have already been mentioned in the preceding outline.

Orthovoltage X-ray Machines

While orthovoltage equipment has become virtually obsolete, it still has an important if limited place in the treatment of such lesions as cancer of the lip, and in intraoral cone therapy in which treatment is given through a small cone inserted into the mouth or other body cavity. Furthermore, modern megavoltage therapy has its roots in the experience gained with orthovoltage and teleradium units over a period of many years.

By orthovoltage x rays we mean those generated at a peak energy ranging from about 140 to 300 kV. The range of 50 to 120 kV applies to superficial therapy of skin or mouth lesions. In both instances, x rays are produced by processes that fulfill the following requirements:

Separation of electrons (space charge). In a conventional x-ray tube (see Figure 17.01) an electric current heats a filament to release an electron cloud or space charge. This remains near the filament until acceleration of space charge electrons occurs.

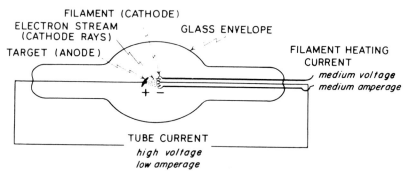

Figure 17.01. Simplified model of an x-ray tube. Note the two circuits: one a filament circuit to heat the filament and liberate electrons, and the other a tube circuit across which the high voltage drives the electrons to the target where their kinetic energy is converted to heat and x rays.

Acceleration of space charge electrons by a high voltage (kilovolts) applied between the cathode and anode. The high voltage drives the electrons toward the anode where they strike the target. Note that there are two circuits in the x-ray tube, one to heat the filament and the other to accelerate the electrons. Also note that the high voltage is so applied that the filament acquires a negative charge (cathode) and the target a positive charge (anode).

Focusing of the electron stream (cathode rays). A concave depression ("cup") in the metal in which the filament is seated acquires a negative charge, thereby focusing the electron stream on a predetermined area of the target.

Sudden slowing of electrons. Upon colliding with the target—usually tungsten—the electrons undergo rapid deceleration (slowing) and, in doing so, produce x rays. The latter arise through *two* processes. First, electrons approaching the nuclei of the target atoms experience a change in velocity on account of the attractive force of the nuclear positive charge. Whenever a fast-moving electron undergoes a change in velocity, it radiates energy in the form of an x-ray photon. Such x-ray production resulting from a change in an electron's velocity is known as *bremsstrahlung* (braking radiation) as shown in Figure 17.02. *Second*, some electrons, on entering the

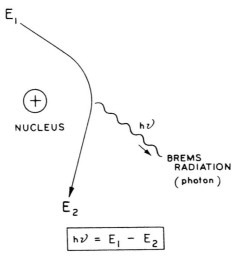

Figure 17.02. Production of *bremsstrahlen* (brems radiation). A fast-moving electron with energy E_1, on approaching an atomic nucleus, experiences an attractive electrostatic force, which changes the direction of the electron along a curved path as shown. In changing direction, the electron radiates energy, hv, which must come from the electron's kinetic energy. Therefore, the electron now moves on with less energy, E_2. The radiated photon (brems ray) *energy* $= E_1 - E_2$.

target atoms, interact with orbital electrons, raising them to higher energy levels or ejecting them completely from the atom. An electron displaced in this way leaves a "hole" in the shell; when this hole is filled by an electron transition from a higher energy level (shell) a characteristic x-ray photon is emitted (see Figure 17.03). Thus, the x-ray beam emitted by the target consists of brems and characteristic x-ray photons. While these photons are individually indistinguishable from each other, they differ in that brems radiation has a continuous spectrum (energy range) from a minimum to a maximum (see Figure 17.04), whereas characteristic

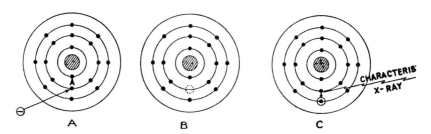

Figure 17.03. Production of characteristic radiation. In *A*, the *electron* in the x-ray tube collides with an inner shell electron in a target atom (actually, they interact by virtue of their electric fields).

In *B*, the atom is in an "excited" state inasmuch as electron *A* has been removed by the "collision."

In *C*, an electron drops from an outer shell to fill the "hole" left by electron *A*; this is accompanied by *characteristic radiation* as the atom returns to its normal state.

radiation exhibits discontinuity in that the photons have limited, discrete energies unique not only for the target element but also for the energy levels between which electron transitions have occurred. Because of the fluctuating kilovoltage, from minimum to maximum (current wave form) in the x-ray tube, as well as production of brems and characteristic radiation, an ordinary x-ray beam is *polyenergetic* or *heterogeneous*, consisting of photons having a variety of energies (and wavelengths). During the collision of electrons with target atoms, the latter are caused to vibrate; this gives rise to a marked heating effect on the anode.

Megavoltage X-ray Machines

As already mentioned, megavoltage x-ray equipment includes two main types: (1) linear accelerator (linac) and (2) betatron. Although tube design and method of electron acceleration in megavoltage machines differ substantially from that in orthovoltage machines, the basic principles of x-ray productions remain the same, namely, development of high speed electrons that strike a suitable target.

Linear Accelerator (linac)

Designed to generate x rays of very high energy, linacs are at present available with energy ratings of 6, 18, and 25 MV (see

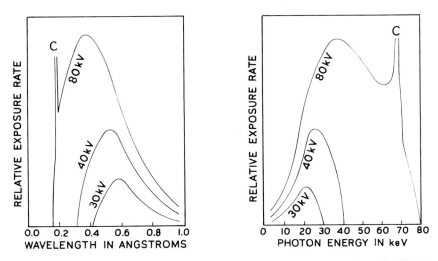

Figure 17.04. Spectral distribution curves of x radiation at 30, 40, and 100 kV. Characteristic radiation appears as a peak at *C* (actually, a group of closely spaced peaks) requiring a minimal potential of 69 kV. The remaining curves represent general radiation. Note that the curves in the left figure are based on wavelength of x rays, whereas those in the right figure are based on photon energy.

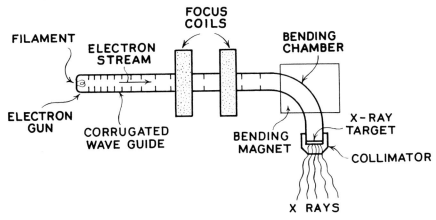

Figure 17.05. Schematic view of a linear accelerator. An electron "gun" injects electrons into the radiofrequency wave guide, which has been set up by a high-frequency alternating current. The electron stream is focused by electromagnetic fields (focus coils) and bent by a magnet so as to approach the target where a high-energy x-ray beam is produced.

Figure 17.05). Electrons are fired into one end of a tube measuring one or two meters in length and undergo acceleration by a standing electromagnetic wave—a radiofrequency wave—across a series

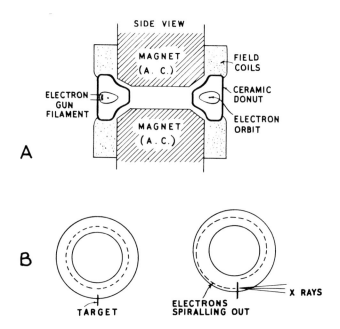

Figure 17.06. Diagram of a betatron in section. *A* = side view. *B* = top view of donut accelerating chamber.

of resonant cavities. Finally, the electrons strike a transmission type tungsten target that emits x rays. An "extractor" makes it possible, if desired, to remove the electrons before they reach the target, for use in *electron beam therapy*.

Betatron

Invented by Kerst in 1941, this has evolved into an x-ray machine generating very high energy beams, usually 22 to 24 MV, for radiotherapy. It also provides high energy electron beams. The betatron consists essentially of a ceramic or glass hollow doughnut-shaped tube placed between the poles of a powerful electromagnet operating at 180 Hz (hertz, cycles per sec). Figure 17.06 shows a simplified diagram of a betatron. Electrons fired into the dough-nut by an electron gun are accelerated by the rapidly changing magnetic field in much the same manner as the electrons in the secondary coil of a transformer. A special device called an *electron peeler* can bring the electrons out of the betatron in the form of an electron beam.

Radionuclides (Gamma-ray Beams)

Cobalt 60 Therapy Unit

Although radium had been used in early teletherapy units, it never became popular because of the insufficient world supply of this rare element, its low radiation output, and the mandatory large source size with its associated large penumbra. When large amounts of cobalt 60 became available, radium was abandoned as a teletherapy source.

Cobalt 60 (^{60}Co). Several properties of ^{60}Co make it eminently useful in *telecurie therapy* (i.e., teletherapy with radionuclides). ^{60}Co is readily manufactured in a nuclear reactor by prolonged irradiation of stable cobalt 59 by slow neutrons according to this equation:

$$\begin{smallmatrix}59\\27\end{smallmatrix}Co + \begin{smallmatrix}1\\0\end{smallmatrix}n \rightarrow \begin{smallmatrix}60\\27\end{smallmatrix}Co + \gamma$$

This is an efficient reaction because cobalt has a strong tendency to capture slow neutrons. Furthermore, after ^{60}Co decays to nickel, two gamma rays are emitted:

$$\begin{smallmatrix}60\\27\end{smallmatrix}Co \rightarrow \begin{smallmatrix}60\\28\end{smallmatrix}Ni + \underset{1.17\,MeV}{\gamma_1} + \underset{1.33\,MeV}{\gamma_2} + \begin{smallmatrix}0\\-1\end{smallmatrix}\beta + \underset{antineutrino}{\nu}$$

Note the average energy 1.25 MeV for the gamma rays (the energy of therapy beams will receive more complete treatment in later sections). The beta particles are removed by the steel capsule enclosing the cobalt source.

Cobalt Source and Housing. A ^{60}Co source has a cylindrical shape and measures 1.5 or 3 cm in diameter by 3 or 4 cm in height. It consists of a leakproof double-walled stainless steel capsule containing the ^{60}Co in one of the following forms: solid slug, pellets, or stacked wafers. The *housing* (see Figure 17.07) consists of a heavy metal, principally lead, which may be alloyed with tungsten. It must have sufficient thickness to reduce the radiation to the permissible level in all directions. Thus, the measured exposure rate at various points one meter from the housing surface shall in no case exceed 10 mR/hr, and the average of all the measurements shall not exceed 2 mR/hr.

Two types of mechanisms are available to move the source into

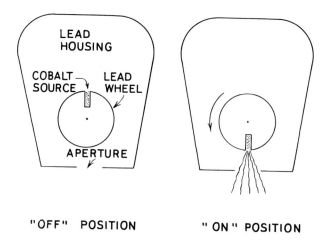

"OFF" POSITION   " ON " POSITION

Figure 17.07. One type of housing for a cobalt 60 source for teletherapy. The cobalt source is housed in a well in the lead wheel. In the "off" position shown on the left, the source points upward, away from the patient. When the timer is activated, a stall-type motor turns the wheel (also tensioning a spring) to the "on" position so the source now points toward the aperture (and patient). At the end of the pretimed treatment, the power to the motor turns off and the spring returns the wheel to the "off" position. (In another type of housing, the source is set in a drawer which is moved by remote control to bring the source over the aperture for the "on" position, and at the end of the treatment, returned to the parked or "off" position.)

position for therapy, in other words to turn the machine "on" and "off." Recall that a radionuclide radiates energy continuously so the radiation cannot actually be turned off. Instead, the source is parked away from the aperture in the housing for the "off" position, and mechanically brought to the aperture for the "on" position. In one system, the ^{60}Co source is mounted in a well in the edge of a solid lead wheel which, in the "off" position, is so turned that the source points upward, away from the aperture. In the "on" position, a stall motor turns the wheel so the source faces the aperture (see Figure 17.07). A spring returns the wheel to the "off" position on completion of the treatment. The other method uses a sliding drawer which advances the source to the aperture by action of a motor, and returns to the remote, protected position by means of a spring.

Cobalt 60 Source Characteristics. Source activity is ordinarily rated in curies (Ci), or some convenient fraction of a curie, usually millicuries (mCi) for brachytherapy. One curie denotes 37 million nuclear transformations (disintegrations) per second, abbreviated $3.7 (10)^7$ dps. Therefore, one millicurie is 37,000 or $3.7 (10)^4$ dps. In the new S.I. units, the becquerel (Bq) is the unit of activity, representing one disintegration per second, so that $1 \text{ Ci} = 3.7 (10)^7$ Bq. Another way of rating a telecobalt source is in terms of its radiation output—roentgens per hour at one meter (RHM). Actual output depends not only on the curiage (number of curies) of the source, but also on the self-absorption of gamma rays in the source itself; this may amount to as much as 20 to 25 percent per cm of source thickness.

Because of protection requirements, cobalt units must be rated according to the maximum activity of the ^{60}Co they can house. For example, a 9000 RHM unit can house a 9000 RHM source according to presently accepted permissible limits, whereas a more active source would require a correspondingly larger housing.

Note that the gamma-ray energy averages 1.25 MeV for all ^{60}Co sources regardless of their activity. However, their gamma-ray output increases as the activity increases. Thus, a 9000 rhm source would have an output twice that of a 4500 rhm source. Obviously, a more active source will permit longer treatment distances in conveniently shorter treatment times than will a less active source.

One of the major drawbacks of ^{60}Co is its continuous decay (5.3-year half-life), requiring periodic adjustment of treatment times at intervals not to exceed three months and necessitating source replacement every five years or so to maintain reasonably short treatment times. On the other hand, once the source is installed, maintenance costs and time lost through equipment failure are negligible.

The steel capsule effectively removes the beta particles, which would otherwise produce an undesirable skin reaction. Filters are not needed to harden the beam (i.e., increase its average penetrating ability) because of the uniform energy of gamma rays. However, beam-flattening filters may be used to make the exposure rate more uniform across the face of the beam.

A multivaned collimator (see Figure 17.08) originally designed by Johns serves not only to delimit the gamma-ray beam but also to reduce penumbra associated with the broad source (compare with collimators in radiography).

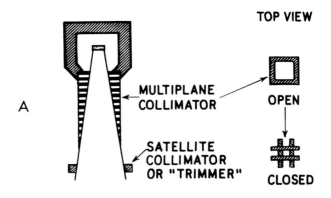

TOP VIEW

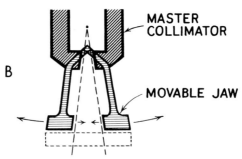

Figure 17.08. In *A* is shown a multivane collimator (Johns' type used in many telecobalt units). In *B* is shown a movable jaw collimator now used in linear accelerators; only one pair is included, the other pair being at a right angle as indicated by the dotted lines.

Cesium 137 Therapy Unit

Because of its relatively long half-life—30 years—we would expect cesium 137 (^{137}Cs) to be an ideal source for teletherapy. However, as we shall see, it has other properties which make it much less suitable than ^{60}Co for this purpose.

Cesium 137, one of the waste products of a nuclear reactor, decays with the emission of two particles and 0.66 MeV gamma ray:

$$_{55}^{137}Cs \rightarrow {}_{56}^{137}Ba + \beta_1 + \beta_2 + \underset{0.66\,MeV}{\gamma} + \underset{antineutrino}{\tilde{\nu}}$$

As with ^{60}Co, a steel capsule enclosing the source removes the beta particles and transmits the gamma rays for teletherapy.

The advantages of ^{137}Cs over ^{60}Co include:

1. Lower cost because it is a waste product.
2. Lower energy gamma ray (0.66 vs 1.25 MeV), allowing smaller source housing and lighter protective wall barrier.
3. Longer half-life.
4. Smaller exit dose (i.e., less radiation to skin on opposite side of body).

However, there are serious disadvantages with ^{137}Cs, including:

1. Low specific activity, that is, low radiation output per gram of material.
2. High degree of self-absorption of gamma rays by the source, about 40 percent.
3. Because of (1) and (2), the source must have a large diameter, usually about 3 cm, producing a larger penumbra.
4. Small specific gamma ray constant, that is, low R/min output per Ci-hr.

Thus, a 1500 Ci source would deliver an exposure rate of about 10 R/min at 70 cm treatment distance. By comparison, a ^{60}Co source having the same activity would deliver about 50 R/min at 70 cm. In practice, the treatment distance with Cs units ranges from 15 to 35 cm to achieve a reasonably short treatment time. Its use is therefore limited to small structures such as the neck.

Because of these disadvantages, ^{137}Cs therapy units have failed to gain a significant place in teletherapy, although they still have limited usefulness in treatment of lesions at intermediate depths, as in the head and neck.

Heavy Particle Generators

Because a considerable number of malignant tumors remain incurable with photon beams for reasons we have already explained, interest has turned in the last few years to heavy particles, principally fast neutrons and negative pi mesons (π^-). While this does not presently concern the general radiologic technologist, it seems appropriate to mention the equipment used to generate these heavy particles for radiotherapy.

Neutron Beam

One of the major sources of neutrons for external beam therapy is the cyclotron, a specialized heavy particle accelerator. In the cyclotron, deuterons (positively charged deuterium nuclei) undergo acceleration in D-shaped chambers called dees (see Figure 17.09) by a radiofrequency low-voltage alternating current.

CYCLOTRON
"DEES"

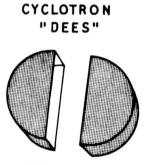

Figure 17.09. Split chambers, called *dees*, in which electrons undergo acceleration in a cyclotron.

The deuterons travel in a flat spiral path controlled by a constant magnetic field (see Figure 17.10). Every time deuterons cross the gap between the dees they receive a voltage "kick," eventually acquiring an extremely high speed with an energy of 15 to 50 MeV. Aiming such high energy deuterons at a beryllium target

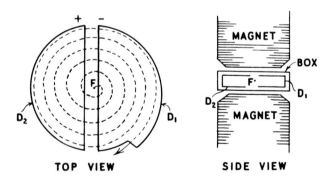

Figure 17.10. Diagram showing the position of the dees (D_1 and D_2) between the magnet poles in a cyclotron.

causes the emission of a concentrated beam of neutrons. Beam collimation is achieved by incorporating steel in the collimator wall to absorb fast neutrons, and surrounding the collimator with a boron and plastic shield to absorb slow neutrons.

Negative Pi Meson (Pion) Beam

The most recent heavy particle modality to be considered for beam therapy uses *negative pi mesons* or, in short, negative pions (symbol п⁻). The particles are released when high energy electrons (about 500 MeV) strike a titanium carbide target. The application of a pion beam in therapy depends on a unique interaction with matter: a pion enters an atomic orbit and is eventually captured by the nucleus, which then disintegrates with a burst of energy called *star formation* (see Figure 17.11). This energy, in turn, releases heavy particle radiation consisting of alpha particles, protons, neutrons, and other nuclear fragments, all of which produce dense ionization (high LET) and severe injury to cells.

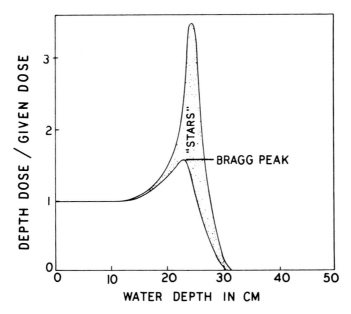

Figure 17.11. Ratio *depth dose/given dose* for a negative pion beam in water. The Bragg peak, resulting from increased ionization as pions slow down, is lower than the peak from star formation. Ideally, the peak region should encompass the tumor volume and an estimated uninvolved layer around the tumor.

Confining such energy release within a tumor improves the chance of its eradication as compared with high energy photons, for reasons discussed earlier (see pages 111–113, 118, 124–125).

BRACHYTHERAPY

We have already defined brachytherapy as irradiation at a short distance. It may be conveniently subdivided into (1) interstitial, (2) intracavitary, and (3) surface therapy. Although the general technologist is rarely if ever involved in brachytherapy, this modality is important to the therapy technologist who may be involved in the procedure.

Interstitial Therapy

Radioactive sources may be implanted interstitially, that is, directly into and near an accessible tumor in a predetermined pattern. Most commonly used are radium 226, cesium 137, gold 198 (seeds), cobalt 60, tantalum 182 (wire), iridium 192 (wire), and californium 252. These will be described as concisely as possible.

Radium

A naturally occurring radionuclide, radium was discovered by Marie and Pierre Curie in 1898. Because of its long half-life, 1622 years, radium used in implants must be removable, which means that such implants are left in position until the calculated dose has been delivered, and then removed.

	URANIUM	RADIUM	RADON	RADIUM A	RADIUM B	RADIUM C	LEAD
				POLONIUM	LEAD	BISMUTH	
Atomic No.	92	88	86	84	82	83	82
Mass No.	238	226	222	218	214	214	206
Half Life	4.5×10⁹yr	1622 yr	3.8 da	3 min	26.8 min	19.7 min	stable

arrows: alpha (URANIUM→RADIUM), alpha (RADIUM→RADON), alpha (RADON→RADIUM A), beta gamma (RADIUM A→RADIUM B), alpha beta gamma (RADIUM B→RADIUM C)

Figure 17.12. The radium transformation (decay) series, with successive product nuclides shown from left to right. This series is the last part of the uranium series. Data on RaD, RaE, and RaF have been omitted.

Radium undergoes radioactive transformation ("decay") to the heavy radioactive gas *radon*, which then decays through a series of daughter nuclides ending up as lead (see Figure 17.12). The decay products as a whole comprise the *radium series*. As ordinarily used, the radium in the form of a salt, radium chloride, is enclosed in a sealed platinum-iridium needle or tube. At the end of one month after manufacture, all of the decay products are in radioactive equilibrium, which means that in a given time interval, say, one second, the same number of atoms of a particular daughter nuclide are appearing (by decay of its parent) as are disappearing (by decay to its daughter). Various members decay by emission of alpha particles, beta particles, and gamma rays. Because of their poor penetrating ability, the alpha and beta particles are absorbed in the wall of the needle, which usually consists of a platinum-iridium alloy 0.5 mm thick. This leaves the gamma rays, which undergo very little absorption and pass into the surrounding tissues. Having an average energy of 0.88 MeV, the gamma rays of the radium series produce about the same effect in tissue as ^{60}Co gamma rays of equal dosage.

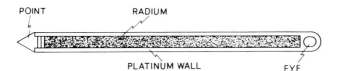

Figure 17.13. Radium needle. The wall is actually a platinum-iridium alloy. Sealed within the needle is the radium in the form of radium chloride. A radium tube has the same general construction except that both ends are rounded. Artificial radionuclides such as cesium 137, iridium 192, gold 198, and tantalum 182 are gradually replacing radium in brachytherapy.

Radium *needles* have an eye at one end and a point at the other (see Figure 17.13) and are available in several standard sizes. High intensity needles have a length of 1 cm, and contain 10 mg radium. Low intensity needles have a loading of 0.66 mg per cm and come in 1, 2, 3, and 4 mg sizes. Radium-containing *tubes* are blunt at both ends.

Distribution of radium needles in an implant must follow some predetermined plan such as the Manchester (Paterson-Parker), the Quimby, or the Martin System. With these systems, one can determine the dosage at various points in and near the tumor, preferably with the aid of a computer. It must be emphasized that the various systems do not give identical dosage distributions.

Radon

As we have just indicated, radon arises by decay of its parent radium:

$$\underset{88}{^{226}}Ra \;\rightarrow\; \underset{86}{^{222}}Rn \;+\; ^{4}_{2}He^{++}$$

radon   alpha particle

Radon, a heavy colorless gas, decays to polonium by emission of an alpha particle. Small, hollow gold seeds containing radon and measuring about 0.75×3 mm with a wall thickness of 0.3 mm can be implanted permanently in and around an accessible tumor. The radon decays rapidly with a half-life of 3.8 days. (Recall that half-life is the time it takes for a radionuclide to decay to one-half its initial activity.) Because of this high rate of decay, an initial loading of, say, 1 mCi of *radon* gives a smaller dose than 1 mCi of *radium* left in the tissues for the same length of time. In fact, 1 mCi of radon remaining permanently in the tissue is equivalent to 133 milligram-hours of radium (i.e., 1 mg Ra left in place for 133 hours). Therefore the amount of radon used in an implant requires adjustment for decay over its lifetime.

Because of the safety problems connected with the manufacture of radon, the danger of leakage from the seeds, and the persistent emission of beta particles from the radon decay products for many years, the use of radon in this country has become obsolete. For these reasons, artificial radionuclides have replaced radon in brachytherapy.

Radium Substitutes

A number of radionuclides have certain advantages over radium for brachytherapy. These include availability, greater safety (no danger of leakage), and a larger variety of available physical forms for special situations. The more important radionuclide substitutes for radium and radon will now be discussed briefly.

Gold 198 (^{198}Au). Available in the form of solid "seeds," ^{198}Au has proven useful in permanent implants similar to radon. The conversion factor from radium must take into account the short half-life of [198]Au — 2.7 days. Thus, we relate the radium exposure in mCi-hr (numerically, closely equivalent to mg-hr for Ra) to the amount of [198]Au "destroyed" (i.e., decaying completely in a permanent implant) as follows:

$$1 \ mCi\text{-}hr \ Ra = 0.038 \ mCi \ \ ^{198}Au \ destroyed, \ or$$
$$1 \ mCi \ \ ^{198}Au \ destroyed = 26.6 \ mCi\text{-}hr \ Ra$$

[198]Au emits a gamma ray with an energy of 0.412 MeV, and a beta particle with an average energy of 0.316 MeV; the latter may be removed by coating each seed with a layer of platinum 0.15 mm thick. Distribution tables, such as those in the Manchester System, require correction for the low energy gamma rays of [198]Au.

Cesium 137 (^{137}Cs). This is becoming one of the more widely used substitutes for radon in *removable* interstitial and intracavitary implants. It is usually prepared as microspheres imbedded in ceramic and encased in stainless steel needles with a 1 mm wall thickness. The conveniently long half-life of [137]Cs — 30 years, with an annual decay factor of 2 percent per year — obviates the frequent decay corrections needed with relatively short-lived nuclides such as [60]Co. The exposure conversion factors are

$$1 \ mCi\text{-}hr \ Ra = 2.55 \ mCi\text{-}hr \ \ ^{137}Cs, \ or$$
$$1 \ mCi\text{-}hr \ \ ^{137}Cs = 0.39 \ mCi\text{-}hr \ Ra$$

[137]Cs emits gamma rays with an energy of 0.66 MeV. While their transmission in tissues approximates that of radium gamma rays, the monoenergetic gamma rays of ^{137}Cs, in the absence of higher energy radiation, requires less protective shielding.

Tantalum 182 ([182]Ta). Another substitute for radium, [182]Ta has suitable properties in removable implants for interstitial therapy when prepared in the form of hairpin-shaped wires 0.2 mm thick, ensheathed in 0.1 mm platinum (Pt). This has been useful in treating bladder cancer. Unfortunately, [182]Ta has a short half-life of only 115 days so that prolonged storage is impracticable. The conversion factor relative to radium is

$$1 \ mCi\text{-}hr \ Ra = 1.2 \ mCi\text{-}hr \ \ ^{182}Ta, \ or$$
$$1 \ mCi\text{-}hr \ \ ^{182}Ta = 0.83 \ mCi\text{-}hr \ Ra$$

[182]Ta emits a spectrum of gamma rays ranging from 0.05 to 1.24

MeV, as well as beta particles which are removed by the Pt sheath.

Iodine 125 (^{125}I). With a half life of 60 days, ^{125}I is the longest lived among the 21 radioisotopes of iodine. Its decay to stable tellurium results in photon radiation only (no beta particles) consisting mainly of 27.4-keV x rays and 35.4-kV gamma rays. Such low energy radiation requires much less protection than other radionuclides such as ^{198}Au and radon. Furthermore, its long "shelf" life is another important advantage over ^{198}Au. Iodine 125 has a specific gamma ray constant of 0.6 rad per mCi at one cm.

In the form of seeds, ^{125}I has been used in a number of radiotherapy centers for permanent interstitial implants, particularly in carcinoma of the prostate. At Memorial Sloan-Kettering Cancer Center, for example, treatment of stage B and early stage C prostatic cancer involves an implant with seeds spaced at 1-cm intervals throughout the gland, the total activity being computed by averaging the three dimensions of the gland and multiplying by a factor of five. The activity of each seed would then be the total activity divided by the number of seeds needed. Such an implant delivers a minimum dose of about 16,000 rads in one year.

Iridium 192 (^{192}Ir). Useful as removable implants, ^{192}Ir can be obtained in the form of seeds ensheathed in stainless steel measuring 0.5 × 0.3 mm. The seeds come loaded, 1 cm apart, along removable nylon ribbons for threading into accessible lesions. The exposure conversion factors relative to radium are as follows:

$$1 \; mCi\text{-}hr \; Ra = 1.6 \; mCi\text{-}hr \; {}^{192}Ir, \; or$$
$$1 \; mCi\text{-}hr \; {}^{192}Ir = 0.63 \; mCi\text{-}hr \; Ra$$

^{192}Ir has a half-life of 74 days and emits beta particles that are absorbed in the stainless steel sheaths, and gamma rays with an energy of only 0.34 MeV (340 keV). They require relatively light protective shielding, the half-value layer being 3 mm lead (Pb). Distribution tables such as those in the Manchester System need correction for the low energy gamma rays.

Californium 252 (^{252}Cf). A transuranic element (beyond uranium in the periodic table, therefore, artificial), ^{252}Cf has completely different properties from the other radium substitutes. Decaying by spontaneous fission with a half-life of 2.6 years, 97 percent of the ^{252}Cf nuclei undergo alpha decay, but the alpha particles are absorbed in the needle wall. The remaining 3 percent of the nuclei undergo slow fission, each giving rise to four neutrons and a gamma ray. The average energy of the neutrons is 2 to 4 MeV. A typical needle contains one microgram (1 millionth of a gram) of ^{252}Cf electroplated on a platinum-iridium thin rod enclosed within a cell. Thus, the needle wall, which consists of platinum-iridium, and the cell contain the ^{252}Cf source. Dosage measurement in a phantom indicates a contribution of about 35 percent from gamma rays and 65 percent from neutrons. Although a 1 microgram needle provides a dose rate about equal to that from a 1 mg radium needle of the same active length (2 cm) the effect on the tissues is much more pronounced with ^{252}Cf because of the large neutron component in the radiation.

Systemic Therapy with Radionuclides

At present two radionuclides have demonstrable value in the systemic treatment of certain diseases—iodine 131 and phosphorus 32.

Iodine 131 (^{131}I)

This radionuclide, obtained from the fission products of the nuclear reactor, or by neutron irradiation of tellurium, has a physical half-life of 8.1 days and emits the following radiation:

beta particles—av. energy 0.182 MeV, of which 90 percent have an energy of 0.608 MeV.

gamma rays—in a spectrum, of which 82 percent have an energy of 0.364 MeV.

^{131}I has proved extremely useful in the treatment of hyperthyroidism (overactivity of the thyroid gland) and about 10 to 15 percent of thyroid cancers. The use of ^{131}I in these conditions depends on selective absorption of iodine by the cells of the thyroid gland. The beta particle component of the radiation is mainly responsible for the therapeutic effect by a factor of 10 to 1 relative to the gamma rays. The gamma ray component makes it possible to diagnose disorders of the thyroid gland by external counting and scanning, and to detect metastatic deposits of thyroid cancer.

Phosphorus 32 (^{32}P)

A pure beta emitter, ^{32}P has a half-life of 14.3 days. The beta particles have an average energy of 0.7 MeV (maximum energy 1.7 MeV). ^{32}P is produced commercially in the nuclear reactor by neutron irradiation of sulfur, according to the following reaction:

$$\underset{\substack{stable \\ sulfur}}{^{32}_{16}S} \quad + \quad \underset{neutron}{^{1}_{0}n} \quad \rightarrow \quad \underset{\substack{radioactive \\ phosphorus}}{^{32}_{15}P} \quad + \quad \underset{proton}{^{1}_{1}H}$$

Phosphorus has the peculiar property of concentrating in cell nuclei (important component of DNA, for example). Since nuclei of malignant cells tend to have a larger mass than those of normal cells, they concentrate about 3 to 5 times as much phosphorus. Furthermore, phosphorus concentrates in bone. Therefore ^{32}P has been found useful in treating such conditions as polycythemia vera (overproduction of red blood cells), both from deposition in the nuclei of primitive red cells and in bone.

Formerly, ^{32}P was also used to treat chronic leukemias, but it has been replaced by chemotherapy (and occasionally radiotherapy), which is much better for this purpose.

In the past, ^{32}P was also used systemically for the relief of pain from skeletal metastasis, principally from carcinoma of the prostate and breast. However, results have been equivocal in most cases. External megavoltage therapy has proven much more effective in controlling pain from localized skeletal deposits.

RADIOTHERAPY

Goals of Radiotherapy

As with any other type of treatment, we must have some idea as to the purpose and goals of radiotherapy. In general, irradiation can be used to control a malignant tumor, or to alleviate (lessen)

one or more of its harmful effects. The four major goals of radiotherapy follow.

Curative Radiotherapy. As the term implies, in curative radiotherapy we aim to deliver a cancericidal (cancer-killing) dose of
radiation to an entire tumor without producing irreparable damage to nearby normal tissues or organs. In other words, while the
tumor is destroyed, the normal tissues can make an adequate
structural and functional recovery. Curative therapy implies a
reasonable probability of achieving a cure or long-term control.

Palliative Radiotherapy. An incurable cancer may produce symptoms such as pain, discharge, or bleeding. In favorable cases moderate doses of radiation may alleviate pain, shrink an unsightly or
foul-smelling mass, or reduce bleeding, without causing unac-
ceptable side effects. Sometimes response is so favorable that the
goal may be changed from palliation to cure.

Combined Radiotherapy and Surgery. Since the advent of high
energy radiation beams (cobalt 60; linac; betatron) which permit
the safe delivery of large doses of radiation, it has been found that
a number of tumors can be cured by a combination of surgery with
either pre- or postoperative irradiation. Examples include certain
cancers of the head and neck, breast, and urinary bladder.

Combination Radiotherapy and Chemotherapy. As a rule, small
tumors respond better than large ones to chemotherapy, just as to
radiotherapy. Therefore we often try to induce as much tumor
shrinkage as possible by irradiation before starting chemotherapy.
Not infrequently, this sequence achieves cancer control. In some
cases, surgery is also included to improve the results even further.

The Planning Process

An acceptable method of irradiation therapy should deliver a
uniform and adequate dose of radiation to a tumor, encompassing
also an adjacent zone of apparently normal tissue into which the
tumor may have penetrated. At the same time, normal tissues
should have a low probability of sustaining serious radiation damage. To accomplish this goal we must proceed with a definite plan
which includes (1) tumor localization, (2) field selection, (3) accu-

rate beam direction, and (4) delivery of an adequate radiation dose in an appropriate number of fractions. As indicated before, careful tumor staging and patient evaluation must be carried out before treatment is planned.

We shall limit our discussion to a few general principles because detailed planning varies widely, depending on the complexity of the particular therapy problem. First, definitions of some of the pertinent terms will be given. These will be followed by a brief discussion of the planning process.

Terminology in Radiotherapy

Exposure in Air. Based on ionization in air, exposure is measured by means of a suitable device such as a Victoreen R-meter, integrating dosimeter, or solid state electrometer. The familiar unit of exposure, the roentgen (R), will eventually be replaced by the S.I. unit of exposure, 1 coulomb/kg (1 coulomb per kg). Exposure measurements in air are valid only for photon energies up to 3 MeV. The output of a therapy source at a particular distance is expressed in R/min or coulomb/kg-sec.

Given Dose (GD). The maximum absorbed dose where the beam enters the body is the given dose. With orthovoltage radiation the given dose is delivered at the skin surface. With megavoltage radiation there is a buildup of primary electrons as the beam enters the body, reaching a maximum at a depth below the surface, that increases with the energy of the photon beam. Thus, the given dose $-D_{max}$ or 100 percent dose $-$ lies at a depth of 5 mm for ^{60}Co gamma rays, 15mm for 6-MV x rays, and 40 mm for 22-MV x rays. For the ^{60}Co *beam* the given dose is obtained from:

$$D_{max} = exposure\ in\ air \times backscatter\ factor \times f \text{ rads} \qquad (1)$$

in familiar terminology. According to the S.I. system, dose should be stated in grays (1 Gy = 100 rads). Some authors are using the centigray (cGy) which is one one-hundreth of a gray; thus, one rad equals one cGy. Dosage with higher energy beams (4 to 25 MV) is derived from measurements in a water phantom.

The *backscatter factor (BSF)* derives from radiation that scatters back from the body toward the entrance port, and is defined by:

$$BSF = D_{max}/D \qquad (2)$$

where D is the dose in air at the same point as D_{max}. The BSF increases with an increase in the area of the treatment field; with an increase in the thickness of the irradiated part; and with a decrease in the energy of the beam. It can be obtained from appropriate depth dose tables. The term f in equation (1) is the conversion factor from exposure in R to absorbed dose in rads (see also pages 27–29).

Percentage Depth Dose (%DD). As an x or gamma-ray beam passes through the body, the dose rate decreases progressively on account of (a) scatter of radiation out of the beam, (b) absorption of radiation by the tissues, and (c) inverse square law (decreasing dose rate as the beam reaches points within the body farther and farther from the source). Tables or graphs, derived from appropriate measurements in "phantoms" composed of tissue-equivalent material, give the percentage of the given dose that reaches various depths in the body along the central axis of the beam. The concept of percentage depth dose is presented in Figure 17.14, and a sample of such data is shown in Table 17.01. The percentage depth dose increases with increasing beam energy, treatment field area, and source-skin distance; and decreases with increasing tumor depth below the surface. Percentage depth dose is also influenced by the type of tissue being traversed by the beam. The depth dose tables have been obtained initially by measurement in material resembling human soft tissue (approximating water density). Therefore, when a beam passes through less dense material such as air, the actual percentage depth dose will be greater than that shown in the tables. Conversely, the percentage depth dose will be less when the beam traverses dense material such as bone.

Tumor Dose (TD). The absorbed dose in the center of a tumor (along the central axis of the beam) is obtained by multiplying the given dose by the percentage depth dose:

$$TD = GD \times \%DD \qquad (3)$$

Or, if a certain tumor dose is to be delivered, the required given dose may be obtained by rearranging the equation; thus:

$$GD = TD/\%DD \qquad (4)$$

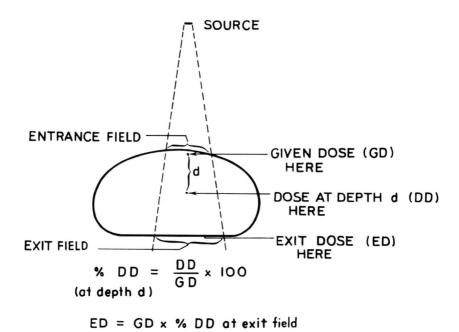

$$\% \ DD = \frac{DD}{GD} \times 100$$

(at depth d)

$$ED = GD \times \% \ DD \ at \ exit \ field$$

Figure 17.14. Concepts of percentage depth dose and exit dose. Dose rate decreases as the beam passes through the body, being less at depth d than at D_{max} where given dose is specified. In terms of percentage, the percent depth dose at a particular depth d equals 100 times the dose at depth d divided by the given dose, as shown. The exit dose (i.e., where the beam leaves the body) equals the given dose times the percent depth dose at that level.

Exit Dose (ED). It must be kept in mind that the beam leaves the body (exits) through a port opposite the entrance field; the former is therefore called the *exit field*. A dose must unavoidably be delivered to the exit field as the beam passes through it, and this is called the *exit dose* (see Figure 17.14). In other words, the exit dose is the absorbed dose in the skin where the beam exits the body. The exit dose may be obtained from

$$ED = GD \times \%DD \ at \ the \ exit \ field \qquad (5)$$

For example, if the part traversed by the beam is 20 cm thick, we look up the %DD in the appropriate table corresponding to a

TABLE 17.01

CENTRAL AXIS PERCENTAGE DEPTH DOSES FOR SQUARE FIELDS

Cobalt 60				Open ended applicator						100 cm SSD
Depth				Field area in cm^2						Depth
in cm	0	16	25	36	49	64	81	100	225	400	in cm
*	100.0	101.5	101.8	102.2	102.5	102.9	103.2	103.5	105.1	106.3	*
											0
0.5**	100.0	100.0	100.0	100.0	100.0	100.0	100.0	100.0	100.0	100.0	0.5
1	95.9	97.1	97.3	97.7	97.9	98.1	98.3	98.6	99.0	98.9	1
2	87.9	91.4	91.9	92.6	92.9	93.3	93.6	93.9	94.6	94.7	2
3	80.7	85.8	86.5	87.5	87.9	88.5	88.9	89.3	90.2	90.5	3
4	73.8	80.2	81.2	82.4	83.0	83.7	84.2	84.7	85.9	86.3	4
5	67.8	74.8	76.0	77.3	78.1	78.9	79.6	80.1	81.6	82.2	5
6	62.3	69.7	70.9	72.4	73.2	74.2	74.9	75.5	77.3	78.1	6
7	57.3	64.8	66.0	67.6	68.4	69.5	70.2	70.9	73.0	74.0	7
8	52.7	60.1	61.3	62.9	63.8	64.9	65.6	66.4	68.7	70.0	8
9	48.5	55.7	56.9	58.4	59.4	60.5	61.2	62.0	64.5	66.1	9
10	44.7	51.5	52.7	54.2	55.2	56.3	57.0	57.8	60.6	62.3	10
11	41.2	47.7	48.8	50.3	51.3	52.4	53.1	53.9	56.9	58.7	11
12	38.0	44.1	45.2	46.7	47.7	48.7	49.5	50.3	53.4	55.3	12
13	35.0	40.8	41.9	43.3	44.3	45.4	46.1	47.0	50.2	52.1	13
14	32.2	37.8	38.9	40.2	41.2	42.3	43.0	43.9	47.1	49.1	14
15	29.6	35.0	36.1	37.4	38.3	39.4	40.1	41.0	44.2	46.2	15
16	27.2	32.5	33.5	34.8	35.6	36.7	37.4	38.3	41.5	43.5	16
17	25.0	30.1	31.1	32.3	33.1	34.2	34.9	35.8	39.0	41.0	17
18	23.0	27.9	28.8	30.0	30.8	31.9	32.6	33.5	36.7	38.6	18
19	21.2	25.8	26.7	27.9	28.7	29.7	30.5	31.3	34.5	36.4	19
20	19.5	23.8	24.7	25.9	26.7	27.7	28.5	29.3	32.4	34.4	20

*This line gives the dose at maximum (5 mm depth), for 100 rads to a small mass of tissue (equilibrium thickness) at the same location in "free space." Division by 100 gives the backscatter factor.

**Dose normalized to 100% at 0.5 cm (depth of maximum electron buildup).

(*Adapted from* Johns, EH, and Cunningham, JR. *The Physics of Radiology*, 1974. Courtesy of Charles C Thomas, Publisher, Springfield, Illinois.)

depth of 20 cm. Suppose this is 30 percent. If the GD is 10 Gy (1000 rads), the exit dose is

$$10 \; Gy \times 0.30 = 3 \; Gy$$
$$or \; 1000 \; rads \times 0.30 = 300 \; rads$$

Field Dose (FD). When a tumor is treated by parallel opposed
fields as in Figure 17.15 such that the entrance and exit fields are
coaxial (i.e., directly opposite to each other), the GD and ED must
be added to obtain the total dose to that field. This sum, then,

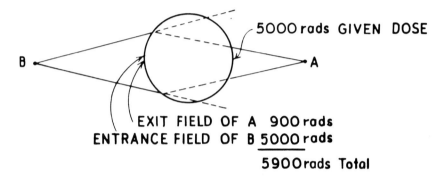

5000 rads GIVEN DOSE

EXIT FIELD OF A 900 rads
ENTRANCE FIELD OF B 5000 rads

5900 rads Total

Figure 17.15. Parallel opposed treatment fields. The exit dose must be added to
the given dose at the coaxial fields to obtain the total fields dose, an extremely
important step in radiotherapy planning to avoid excessive dosage at the D_{max}
level. As shown in Figure 17.17, isodose curves reveal the dosage distribution at
all levels in the treatment field.

becomes the *total field dose*. In Figure 17.16 the GD and ED for field
A, when added together, give the field dose (FD) for field A; and
similarly for field B. This is extremely important because serious
overdosage of skin could occur if the contribution to it from the
exit dose were to be ignored.

Source-Skin Distance (SSD). This is the *distance, along the central
axis of the beam, from the radiation source to the entrance skin surface.*
It applies to all radiotherapy beams. However, the *calibration distance* is the distance from the source to the site of D_{max} (i.e., level of
maximum buildup or dose). Thus, we calibrate an orthovoltage
beam at the SSD because D_{max} is at the skin surface. But a ^{60}Co
gamma-ray beam with an 80-cm SSD is actually calibrated at a
distance of 80 cm + 0.5 cm = 80.5 cm because D_{max} in this case lies
0.5 cm below the skin surface. For a 6-MV x-ray beam at a 100-cm
SSD the calibration distance is 100 cm + 1.5 cm = 101.5 cm; and
for a 25-MV x-ray beam, 100 cm + 4 cm = 104 cm.

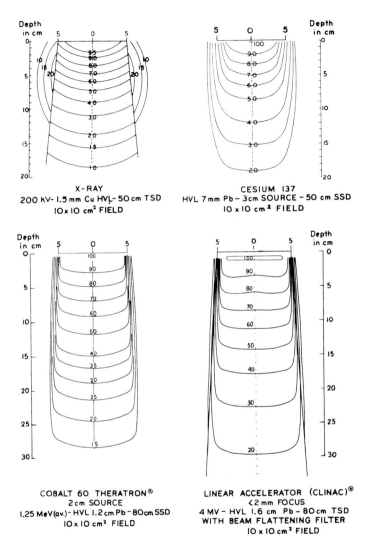

Figure 17.16. Isodose curves for various teletherapy units under typical operating conditions. Note the greater percent depth dose at a particular depth with increasing beam energy. Although the 4-MV x-ray beam is more sharply defined at the entrance, this difference fades at increasing depths in the body. (Curves for orthovoltage x rays adapted from Johns HE, Cunningham JR; cesium 137 from Picker Medical Products Division; cobalt 60 from Atomic Energy of Canada; Clinac from Varian.)

Isodose Curves. We learned above that percentage depth dose refers to the fraction of a given dose that reaches a particular depth in the body along the central axis of the beam. However, we must also know the distribution of radiation at other points in the beam because dosage is not uniform across the beam at any given depth. In other words, as shown in Figure 17.16, the percent depth dose at a particular depth is smaller toward the periphery of the beam than at the center. In planning a course of radiotherapy we must take such isodose distributions into account to avoid regions of over- or undertreatment. Isodose curves are really map lines corresponding to points of equal dosage (*iso* same). For example, in Figure 17.16 the 40 percent line indicates all the points within the irradiated volume that receive a 40 percent depth dose. Figure 17.17 shows how isodose curves appear in a simple treatment plan with parallel opposed fields.

On the basis of tumor type and extent, as well as the tolerance of the normal tissues included in the irradiated volume, the radiation therapist selects the appropriate dose and fractionation schedule. Care must be taken to shield vital organs to avoid exceeding their tolerance dose. For example, the kidneys must be shielded by lead of sufficient thickness to limit their dose to 2200 rads (22 Gy) with conventional fractionation; increasing the dose above this level, or reducing the number of fractions in the same overall treatment time causes a sharp rise in the probability of serious radiation injury to the kidney.

The dosimetrist plans the delivery of the radiation dose to the tumor. This includes such factors as the selection of fixed or moving (rotational) beams. With fixed beam modality, two or more cross-firing beams aimed at the tumor from various directions serve to spare normal tissue and provide more uniform irradiation of the tumor. With moving beam therapy, the beam rotates around the patient over a preselected arc; it actually represents a summation of many fixed beams.

Accurate planning requires a *simulator*, which duplicates the geometric factors of the therapy beam such as source diameter, SSD or TSD, tumor site and size, and beam direction, but uses a radiographic beam. The tumor is visualized directly or with a contrast medium, by fluoroscopy and radiography with the simu-

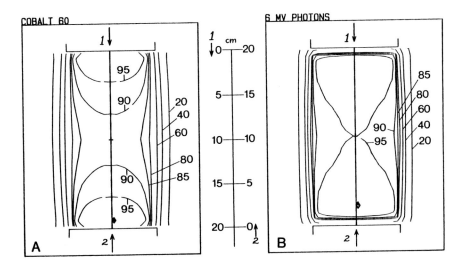

Figure 17.17. *A.* Isodose distribution with parallel opposed fields, cobalt 60 beam. With this beam energy (1.25 MeV) a 100-rad given dose to each field delivers a maximum dose of 128 rads at the small diamond. The central volume receives 80 to 90 percent of the maximum dose, or 102 to 115 rads.

B. Isodose distribution with parallel opposed fields, 6-MV x-ray beam. For a 100-rad given dose to each field, the maximum dose is 144 rads at the small diamond. The relatively uniform dose to the central volume is now 90 to 95 percent of the maximum dose, or 130 to 137 rads. Compare this with the above data for the cobalt 60 beam. (Isodose curves by courtesy of Frederick Hager, Physicist, East Texas Cancer Center, Tyler, Texas.)

lator, and the entrance ports for the treatment beams are marked on the skin with special dye or paint. Beam direction with the therapy beam must exactly duplicate that with the simulator to assure complete tumor coverage.

The dosimetrist then plans the treatment details with regard to dosage to all parts of the tumor and neighboring tissues. In recent years, computerized tomography (CT) has contributed greatly to the exactness of treatment planning. Finally, adequate coverage of the tumor is verified by radiographs obtained through each port with the actual therapy beam. The radiotherapist reviews the entire plan before putting it into effect.

In some cases, *shrinking field therapy* is used—since the periphery of the tumor is better oxygenated and more radiosensitive than the center, port size may be reduced as the tumor shrinks, toward the end of the treatment course. However, this requires superb judgment on the radiotherapist's part to avoid under-treatment of the outer regions of the tumor.

Under certain conditions, such as poor health of the patient, a treatment program may be divided by a two- or three-week rest period about midway through the course. This is known as *split-course* therapy. It may contribute to improved tolerance of the patient and to a better response of the tumor because of increased opportunity for reoxygenation.

Observation of the Patient

It is extremely important that as a course of therapy progresses, the patient be closely watched for untoward reactions, especially when radiosensitive normal organs such as the bowel lie within the beam. The technologist must be continually alert to the patient's complaints or reactions and report them promptly to the radiotherapist. In some instances the complaints may reflect a serious condition not related to therapy; for example, a patient who develops chest pain may minimize the symptoms and ascribe them to the radiation therapy, whereas the underlying condition may actually be a heart attack. It therefore behooves the radiotherapist to be readily available to evaluate and deal with any complaints or other problems that may arise during the course of therapy.

REFERENCES

Adelstein SJ, Dealy JB. Hematologic responses to human whole body irradiation. *Am J Roentgenol 93*:927, 1965.

Anderson, WAD, Kissane JM (Eds.). *Pathology*, ed. 7. St Louis, Mosby, 1971.

Bailar JC III. Mammography: a contrary view. *Ann Int Med 84*:77, 1976.

Baker DG. Medical radiation exposure and genetic risks. *S Med J 73*:1247, 1980.

Barendsen GW. Responses of cultured cells, tumors, and normal tissues to radiations of different linear energy transfers. In Ebert M, Howard A (Eds): *Current Topics in Radiation Research*. Amsterdam, North Holland Publishing Co, vol IV, 1968.

Baserga R. The cell cycle. *N Engl J Med 304*:453, 1981.

Beahrs OH, Shapiro S, Smart C, *et al.* Report of the working group to review NCI/ACS breast cancer detection and demonstration projects. *J Natl Cancer Inst 62*:641, 1979.

BEIR (Committee on the Biological Effects of Ionizing Radiation.) *The Effects on Populations of Exposure to Low Levels of Ionizing Radiation*. Washington National Academy Press, 1980.

Bergonié J, Tribondeau L. Action of x rays on the testicle of the white rat. *Compt Rend Soc Biol 57*:400, 1904.

Bicher HI, Sandhu TS, Hetzel FW. Clinical thermoradiotherapy. In Henry Ford Hospital Special Symposium: *Clinical Hyperthermia Today*. June 21, 1980.

Boice JD, Land CE. Adult leukemia following diagnostic x ray? *Am J Pub Health 69*:137, 1979.

Boice JD, Monson RR. X-ray exposure and breast cancer. *Am J Epidemiol 104*:349, 1976.

Braestrup CB, Vikterlöff KJ. *Manual on Radiation Protection in Hospitals and General Practice*. Vol I. Geneva, World Health Organization, 1974.

Brent RL, Gorson RO. Radiation exposure in pregnancy. *Current Problems in Radiology*. Vol II, No 2. Chicago, Year Book, 1972.

Brill AB, Tomonaga M, Heysell RM. Leukemia in man following exposure to ionizing radiation. A summary of the findings in Hiroshima and Nagasaki and a comparison with other human experience. *Ann Int Med 56*:590, 1962.

BRH (Bureau of Radiological Health). *Population Exposure to X Rays, U.S.* DHEW(FDA). U.S. Government Printing Office, Washington, DC, 1964.

BRH (Bureau of Radiological Health). *Population Exposure to X Rays, U.S.* DHEW(FDA). U.S. Government Printing Office, Washington, DC, 1970.

BRH (Bureau of Radiological Health. *Gonad Doses and Genetically Significant Dose from Diagnostic Radiology, U.S., 1964–1970.* DHEW(FDA) 76-8034. U.S. Government Printing Office, Washington, DC.

Campbell JA. X-ray pelvimetry: useful procedure or medical nonsense. *J Natl Med Assoc 68*:514, 1976.

Canti RG, Spear FG. The effect of gamma radiation on cell division in tissue culture *in vitro*. *Part II. Proc R Soc London (Biol) 105*:93, 1929.

Chiacchierini RP, Lundin FE, Jr. Benefit/risk ratio of mammography. In *Breast Carcinoma, the Radiologist's Expanded Role*. Logan WW (Ed.). New York, John Wiley & Sons, 1977.

Cohen L. Theoretical "iso-survival" formulae for fractionated radiation therapy. *Brit J Radiol 41*:522, 1968.

Collins VP, Loeffler RK, TiveyH. Observations on growth rates of human tumors. *Am J Roentgenol 76*:988, 1956.

Copenhaver WM, Kelly DE, Wood RL (Eds.). *Bailey's Textbook of Histology*. Baltimore, Williams & Wilkins, 1978.

Court-Brown WM, Doll R. Mortality from cancer and other causes after radiotherapy for ankylosing spondylitis. *Brit Med J 2(5474)*:1327, 1965.

Court-Brown WM, Doll R, Hill AB. The incidence of leukemia following exposure to diagnostic radiation *in utero*. *Brit Med J 2*:1539, 1960.

Coutard H. Roentgen therapy of epithelioma of the tonsillar region, hypopharynx, and larynx from 1920 to 1926. *Am J Roentgenol 28*:313, 1932.

Dalrymple GV, Gaulden ME, Kollmorgen GM, Vogel HH (Eds.). *Medical Radiation Biology*. Philadelphia, Saunders, 1973.

Dekaban AS. Abnormalities in children exposed to x radiation during various stages of gestation; tentative time table of radiation to the human fetus. *J Nucl Med Part 1. 9*:471, 1968.

Duncan W, Nias AHW. *Clinical Radiobiology*. New York, Churchill Livingstone, 1977.

Elkind MM, Sutton H. X-ray damage and recovery in mammalian cells in culture. *Nature (London) 184*:1293, 1959.

Elkind MM, Sutton-Gilbert H. Radiation response of mammalian cells grown in culture. I. Repair of x-ray damage in surviving Chinese hamster cells. *Radiat Res 13*:556, 1960.

Elkind MM, Sutton-Gilbert H, Moses WB, Alescio T, Swaim RW. Radiation response of mammalian cells in culture. V. Temperature dependence of the repair of x-ray damage in surviving cells (aerobic and hypoxic). *Radiat Res 25*:359, 1965.

Elkind MM, Whitmore GF. *The Radiobiology of Cultured Mammalian Cells*. New York, Gordon and Breach, 1967.

Ellis F. Nominal standard dose and the ret. *Brit J Radiol 44*:101, 1971.

Errera M, Forssberg A. *Mechanisms in Radiobiology*. New York, Academic Press, 1960, 1961.

Feig SA. Can breast cancer be radiation induced? In *Breast Carcinoma, the Radiologist's Expanded Role*. Logan WW (Ed.). New York, John Wiley, 1977.

Fletcher GH. *Textbook of Radiotherapy*, ed. 3. Philadelphia, Lea & Febiger, 1980.

Fowler JF, Morgan RL, Wood CAP. Pretherapeutic experiments with fast neutron beam from the Medical Research Council cyclotron. I. The biological and physical advantages and problems of neutron therapy. *Brit J Radiol 36*:77, 1963.

Fricke H, Hart EJ. Reactions induced by photoactivation of the water molecule. *J Chem Phys 4*:418, 1936.

Gaulden ME. Possible effects of diagnostic x rays on the human embryo and fetus. *J Ark Med Soc 70*:424, 1974.

Gray JE. Editorial: The radiation hazard—let's put it in perspective. *Mayo Clinic Proc 54*:809, 1979.

Gray LH. The initiation and development of cellular damage by ionizing radiation. *Brit J Radiol 26*:609, 1953.

Hall EJ. Radiobiology of heavy particle radiation therapy: cellular studies. *Radiology 108*:119, 1973.

Hall EJ. *Radiobiology for the Radiologist*, 2nd ed. Hagerstown, Harper & Row, 1978.

Hall EJ, Lehnert S, Roizin-Towle L. Split-dose experiments with hypoxic cells. *Radiology 112*:425, 1974.

Hammer-Jacobsen E. Therapeutic abortion on account of x-ray examination during pregnancy. *Danish Med Bull 6*:113, 1959.

Hempelmann LH. Risk of thyroid neoplasms after irradiation in childhood. *Science 160*:159, 1968.

Hempelmann LH, Hall WJ, Phillips M, *et al.* Neoplasms in persons treated with x rays in infancy. Fourth survey in 20 years. *J Natl Cancer Inst 55*:519, 1975.

Hilaris BS (Ed.). *Handbook of Interstitial Brachytherapy*. Acton, MA, Publishing Sciences Group, 1975.

Howard A, Pelc SR. Synthesis of deoxyribonucleic acid in normal and irradiated cells and its relation to chromosome breakage. *Heredity 6*:261, 1953.

Jans RG, Butler PF, McCrohan JL, Thompson WE. The status of film-screen mammography: results of the BENT study. Radiology 132:197, 1979.

Johns HE, Cunningham JR. *The Physics of Radiology*. Springfield, Thomas, 1978.

Joyce DN, Giwa-Osagie F, Stevenson GW. Role of pelvimetry in active management of labor. *Brit Med J 4*:505, 1975.

Kallman RF. The phenomenon of reoxygenation and its implications for fractionated radiotherapy. *Radiology 105*:135, 1972.

Kaplan HS. *Hodgkin's Disease*. Cambridge, Harvard University Press, 1980.

Kelly KM, Madden DA, Arcarese JS, Barnett M, Brown RF. The utilization and efficacy of pelvimetry. *Am J Roentgenol 125*:66, 1975.

Kneale GW, Stewart AM. Mantel-Haenzel analysis of Oxford data. II. Independent effects of fetal irradiation subfactors. *J Natl Cancer Inst 57*:1009, 1976.

Lacassagne A, Gricouroff G. *Action of Radiation on Tissues. An Introduction to Radiotherapy*. New York, Grune & Stratton, 1958.

Laube DW, Varner MW, Cruikshank DP. A prospective evaluation of x-ray pelvimetry. *JAMA 246*:2187, 1981.

Lea DE. *Actions of Radiations on Living Cells*. New York, Macmillan, 1947.

Lester RG. A radiologist's view of the benefit/risk ratio of mammography. In

Breast Carcinoma, the Radiologist's Expanded Role. Logan WW (Ed.) New York, John Wiley, 1977.

Mackenzie I. Breast cancer following multiple fluoroscopies. *Brit J Cancer* 19:1, 1965.

MacMahon B. Prenatal x-ray exposure and childhood cancer. *J Natl Cancer Inst* 28:1173, 1962.

McFarland W, Pearson HA. Hematologic events as dosimeters in human total-body irradiation. *Radiology 80*:850, 1963.

McKusick VA. *Mendelian Inheritance in Man: Catalogs of Autosomal Dominant, Autosomal Recessive, and X-linked Phenotypes.* Baltimore, The Johns Hopkins University Press, 1978.

Meredith WS, Massey JB. *Fundamental Physics of Radiology.* Bristol (England), John Wright & Sons, 1977.

Moskowitz M, Gartside PS, Gardella L, deGroot I, Guenther D. The breast cancer screening controversy: a perspective. In *Breast Carcinoma, the Radiologist's Expanded Role.* Logan WW (Ed.). New York, John Wiley, 1977.

Murphy DP. *Congenital Malformations.* Philadelphia, Lippincott, 1947.

Myrden JA, Hiltz JE. Breast cancer following multiple fluoroscopies during artificial pneumothorax treatment of pulmonary tuberculosis. *Canad Med Assoc J* 100:1032, 1969.

NCRP Publications. National Council on Radiation Protection and Measurements. 7910 Woodmont Avenue, Washington, DC 20014.

No.   *Title*

33 *Medical X-ray and Gamma-ray Protection for Energies up to 10 MeV — Equipment Design and Use* (1968).

39 *Basic Radiation Protection Criteria* (1971).

43 *Review of the Current State of Radiation Protection Philosophy* (1975).

45 *Natural Background Radiation in the United States* (1975).

48 *Radiation Protection for Medical and Allied Health Personnel* (1976).

53 *Review of NCRP Radiation Dose Limit for Embryo and Fetus in Occupationally Exposed Women* (1977).

54 *Medical Radiation Exposure of Pregnant and Potentially Pregnant Women* (1977).

66 *Mammography* (1980).

68 *Radiation Protection in Pediatric Radiology* (1981).

NCRP Publications. *Quantitative Risk in Standard Setting.* Proceedings of the Sixteenth Annual Meeting as Presented at the National Academy of Sciences (1980).

Paterson R. *The Treatment of Malignant Disease by Radium and X rays.* London, Arnold, 1963.

Pizzarello DJ, Witcofski RL. *Basic Radiation Biology,* 2nd ed. Philadelphia, Lea & Febiger, 1975.

Powers WE, Tolmach IJ. Demonstration of an anoxic component in a mouse tumor cell population by in vivo assay of survival following irradiation. *Radiology 83*:328, 1964.

Puck TT, Marcus PI. Action of x rays on mammalian cells. *J Exp Med 103*:653, 1956.

Puck TT, Marcus PI, Cieciura SJ. Clonal growth of mammalian cells in vitro. *J Exp Med* 103:272, 1956.

Radiation Effects Research Foundation (RERF) quoted in *NCRP Report No. 66*, p 34, 1980.

Rafla S, Rotman M. *Introduction to Radiotherapy.* St Louis, Mosby, 1974.

Robbins SL. *Textbook of Pathology with Clinical Applications.* Philadelphia, Saunders 1962.

Rosenstein M. *Handbook of Selected Organ Doses for Projections Common in Diagnostic Radiology.* U.S. Department of HEW (PHS/FDA) BRH, Rockville, MD, 1976.

Rossi HH. Correlation of radiation quality and biological effect. *Ann NY Acad Sci* 114:4–15, 1964.

Rotblat J, Lindop P. Long-term effects of a single whole-body exposure of mice to ionizing radiations. II. Causes of death. *Proc R Soc London (Biol) 154*:350, 1961.

Rubin P, Casarett GW. *Clinical Radiation Pathology. Vol I & II.* Philadelphia, Saunders, 1968.

Rugh R. Low levels of x irradiation and the early mammalian embryo. *Am J Roentgenol 87*:559, 1962.

Rugh R. The impact of ionizing radiations on the embryo and fetus. *Am J Roentgenol 89*:182, 1963.

Rugh R. X-ray induced teratogenesis in the mouse and its possible significance to man. *Radiology 99*:433, 1971.

Russell LB, Russell WL. An analysis of the changing radiation response of the developing mouse embryo. *J Cell Physiol (Suppl 1) 43*:103, 1954.

Shore RE, Hempelmann LH, Kowaluk E, Mansur PS, Pasternack BS, Albert RE, Haughie GE. Breast neoplasms in women treated with x rays for acute post-partum mastitis. *J Natl Cancer Inst 59*:813, 1977.

Sinclair WK. Cyclic x-ray responses in mammalian cells in vitro. *Radiat Res 33*:620, 1968.

Smith PG, Doll R. *Age- and Time-dependent Changes in the Rates of Radiation-induced Cancers in Patients with Ankylosing Spondylitis Following a Single Course of X-ray Treatment (p 205–218).* International Atomic Energy Report *SM-224/711*, Vienna, 1978.

Strandqvist M. Studies on the cumulative effect of Roentgen rays in fractionation (in German). *Acta Radiol (Suppl 55)*, 1944.

Strax P, Venet L, Shapiro S. Value of mammography in reduction of mortality from breast cancer in mass screenings. *Am J Roentgenol* 112:686, 1973.

Suit H. Consideration of fractionation schedules for radiation dose. *Radiology 105*:51, 1972.

Taliaferro WH, Taliaferro LG, Jaroslav BN. *Radiation and Immune Mechanisms.* New York, Academic Press, 1964.

Terasima T, Tolmach LJ. Variations in several responses of HeLa cells to x radiation during the division cycle. *Biophysical J 3*:11, 1963.

Thoday JM, Read J. Effect of oxygen on the frequency of chromosome aberrations produced by x rays. *Nature 160*:608, 1947.

Thomlinson RH, Gray LH. The histological structure of some human lung cancers and the possible implications for radiotherapy. *Brit J Cancer* 9:539, 1955.

Travis EL. *Primer of Medical Radiobiology.* Chicago, Year Book Medical Publishers, 1975.

Trout ED, Kelley JP, Cathey GA. The use of filters to control radiation exposure to the patient in diagnostic radiology. *Am J Roentgenol* 63:946, 1952.

UNSCEAR. United Nations Scientific Committee on the Effects of Atomic Radiation. General Assembly, 33rd Session. *Sources and Effects of Ionizing Radiation.* Suppl. No. 40 (A/32/40) 1977. United Nations, New York.

Upton AC. The dose response relation in radiation induced cancer. *Cancer Res* 21:717, 1961.

Upton AC. Radiation carcinogenesis. (In *Medical Radiation Biology*, Dalrymple GV, et al., Eds.) p 220. Philadelphia, Saunders, 1973.

Upton AC, Beebe GW, Brown JM, Quimby EH, Shellabarger CJ. Report of NCI ad hoc Working Group on the risks associated with mammography in mass screening for the detection of breast cancer. *J Natl Cancer Inst* 59:481, 1977.

U.S. Department of Health Education and Welfare (FDA). *Gonadal Shielding in Diagnostic Radiology.* DHEW(FDA) 74-8028. Rockville, MD 20852.

U.S. Department of Health and Human Services (FDA). *The Selection of Patients for X-ray Examinations: the Pelvimetry Examination,* 1981.

van Putten LM, Kallman RF. Oxygenation states of a transplantable tumor during fractionated radiotherapy. *J Natl Cancer Inst* 40:441, 1968.

von Essen CF. Roentgen therapy of skin and lip carcinoma: factors influencing success and failure. *Am J Roentgenol* 83:556, 1960.

von Essen CF. A spatial model of time-dose-area relationship in radiation therapy. *Radiology* 81:881, 1963.

Wanebo CK, Johnson KG, Sato K, Thorslund TW. Breast cancer after exposure to the atomic bombings of Hiroshima and Nagasaki. *N Engl J Med* 279:667, 1968.

Watson JD, Crick FHC. The structure of DNA. *Cold Spring Harbor Symposium* 18:123, 1953.

Watson JD. *The Double Helix.* New York, Atheneum Press, 1968.

Withers HR. Biologic basis for high-LET radiotherapy. *Radiology* 108:131, 1973.

Wochos JF, Cameron JR. *Patient Exposure from Diagnostic X rays: an Analysis of 1972-1974 NEXT Data.* DHEW Publication No. (FDA) 77-8020, 1977.

INDEX